Matteo Bianchi

Chronic Granulomatous Disease and Neutrophil Extracellular Traps

Matteo Bianchi

Chronic Granulomatous Disease and Neutrophil Extracellular Traps

CGD gene therapy and anti-Aspergillus activity of reconstituted NET formation

Südwestdeutscher Verlag für Hochschulschriften

Impressum / Imprint
Bibliografische Information der Deutschen Nationalbibliothek: Die Deutsche
Nationalbibliothek verzeichnet diese Publikation in der Deutschen Nationalbibliografie;
detaillierte bibliografische Daten sind im Internet über http://dnb.d-nb.de abrufbar.
Alle in diesem Buch genannten Marken und Produktnamen unterliegen warenzeichen-,
marken- oder patentrechtlichem Schutz bzw. sind Warenzeichen oder eingetragene
Warenzeichen der jeweiligen Inhaber. Die Wiedergabe von Marken, Produktnamen,
Gebrauchsnamen, Handelsnamen, Warenbezeichnungen u.s.w. in diesem Werk berechtigt
auch ohne besondere Kennzeichnung nicht zu der Annahme, dass solche Namen im Sinne
der Warenzeichen- und Markenschutzgesetzgebung als frei zu betrachten wären und
daher von jedermann benutzt werden dürften.

Bibliographic information published by the Deutsche Nationalbibliothek: The Deutsche
Nationalbibliothek lists this publication in the Deutsche Nationalbibliografie; detailed
bibliographic data are available in the Internet at http://dnb.d-nb.de.
Any brand names and product names mentioned in this book are subject to trademark,
brand or patent protection and are trademarks or registered trademarks of their respective
holders. The use of brand names, product names, common names, trade names, product
descriptions etc. even without a particular marking in this works is in no way to be
construed to mean that such names may be regarded as unrestricted in respect of
trademark and brand protection legislation and could thus be used by anyone.

Coverbild / Cover image: www.ingimage.com

Verlag / Publisher:
Südwestdeutscher Verlag für Hochschulschriften
ist ein Imprint der / is a trademark of
AV Akademikerverlag GmbH & Co. KG
Heinrich-Böcking-Str. 6-8, 66121 Saarbrücken, Deutschland / Germany
Email: info@svh-verlag.de

Herstellung: siehe letzte Seite /
Printed at: see last page
ISBN: 978-3-8381-3384-3

Zugl. / Approved by: Zurich, University of Zurich, Switzerland, Diss., 2011

Copyright © 2012 AV Akademikerverlag GmbH & Co. KG
Alle Rechte vorbehalten. / All rights reserved. Saarbrücken 2012

Gene Therapy for Chronic Granulomatous Disease and anti-*Aspergillus* Activity of Reconstituted Neutrophil Extracellular Trap Formation

Dissertation

zur

Erlangung der naturwissenschaftlichen Doktorwürde

(Dr. sc. nat.)

vorgelegt der

Mathematisch-naturwissenschaftlichen Fakultät

der

Universität Zürich

von

Matteo BIANCHI

von

Coldrerio TI

Promotionskomitee

Prof. Dr. Cornel Fraefel (Vorsitz)

Prof. Dr. Alexandra Trkola

PD Dr. med. Janine Reichenbach (Leitung der Dissertation)

Zürich, 2011

Die vorliegende Arbeit wurde von der Mathematisch-naturwissenschaftlichen Fakultät der Universität Zürich im Herbstsemester 2011 als Dissertation angenommen. Promotionskommitee: Prof. Dr. Cornel Fraefel (Vorsitz), Prof. Dr. Michael Hengartner, Prof. Dr. Alexandra Trkola, PD Dr. med. Janine Reichenbach (Leitung der Dissertation).

Table of Contents

Zusammenfassung	V
Summary	VII
Introduction	1
Chapter 1 Restoration of NET Formation by Gene Therapy in CGD controls Aspergillosis	27
Chapter 2 Restoration of anti-*Aspergillus* defense by neutrophil extracellular traps in human chronic granulomatous disease after gene therapy is calprotectin-dependent	47
Chapter 3 *In vitro* models for NADPH oxidase function to evaluate efficacy of new myelospecific γ-retroviral SIN vectors for X-CGD gene therapy	83
Discussion and Outlook	99
Abbreviations	107
Original publications	110
Acknowledgements	135
Curriculum vitae	137

Zusammenfassung

Die septische Granulomatose (engl. chronic granulomatous disease, CGD) ist eine Gruppe von primären Immundefekten, die durch Mutationen in für die Phagozyten Nikotinamid Adenin Dinucleotid Phosphat (NADPH) Oxidase kodierenden Genen (gp91^{phox}, p22^{phox}, p47^{phox}, p67^{phox}, and p40^{phox}) bedingt ist. CGD Phagozyten produzieren keine reaktiven Sauerstoffverbindungen (engl. reactive oxygen species, ROS), töten Mikroben schlecht ab und betroffene Patienten leiden daher an rezidivierenden lebensbedrohlichen Infektionen mit Bakterien und Schimmelpilzen. Aspergillus spp. Infektionen, die Pneumonien und disseminierte Infektionen hervorrufen können, sind die Haupttodesursache von CGD Patienten. Bisher war es unklar, wie gesunde Neutrophile Granulozyten Aspergillus Spezies kontrollieren.

Wir haben daher die Rolle von NADPH Oxidase abhängigen extrazellulären Neutrophilen Fallen (eng. Neutrophil Extracellular Traps, NETs) in Hinsicht auf die anti-Aspergillus Abwehr bei Patienten mit CGD untersucht. NETs werden nach Aktivierung von Neutrophilen freigesetzt und bestehen aus Chromatin (DNA und Histone), das mit granulären und zytoplasmatischen antimikrobiellen Proteinen besetzt ist. NETs binden und halten Bakterien und Pilze gefangen und setzen sie antimikrobiellen Proteinen aus. Die Bildung von NETs benötigt zwingend ROS, die von der NADPH Oxidase produziert werden. Daher können CGD Neutrophile keine NETs bilden. *Wir haben erstmals gezeigt, dass mikrobizide NETs gegen A. nidulans Konidien und Hyphen in vitro wirksam sind, und dass die Rekonstitution der NADPH Oxidase Aktivität durch Gentherapie (GT) bei einem Patienten mit X-chromosomal vererbter CGD die NETs Bildung wiederhergestellt hat.* Dies hat zur Ausheilung einer schweren invasiven A. nidulans Infektion in vivo beigetragen. *Wir haben ferner die kritische Rolle von NET-Assoziiertem Calprotectin demonstriert, das mittels Zink-Sequestrierung Dosis-abhängig eine fungistatische oder fungizide anti-Aspergillus Wirkung hat.*

GT für CGD soll den Defekt der NADPH Oxidase Untereinheiten durch die Insertion einer normalen Kopie des verantwortlichen defekten Gens in autologe hämatopoetischen Stammzellen des jeweiligen Patienten funktionell korrigieren und die Krankheit dauerhaft heilen. In internationaler Kollaboration wurden kürzlich in unserer Abteilung für Immunologie/Hämatologie/KMT am Universitätskinderspital Zürich γ-retrovirale GT Vektoren in klinischen Phase I/II Studien kürzlich ausführlich getestet und die klinische Wirksamkeit der GT für X-chromosomale CGD (gp91^{phox} Defekt) gezeigt. Allerdings machen unerwartete Nebenwirkungen wie Ausschalten (Silencing) der Transgenaktivität und Transaktivierung von Onkogenen eine Verbesserung der Vektoren-Effizienz und -Sicherheit vor künftigen klinischen GT Studien an CGD Patienten zwingend nötig.

Zusammenfassung

Daher haben wir als Kandidaten für künftige klinische X-CGD GT Studien die nächste Generation von γ-retroviralen Selbst-Inaktivierenden (SIN) Vektoren entwickelt und funktionell getestet. Unter den in vitro *getesteten myelospezifischen Vektoren, haben wir die humanen MRP8 und miRNA-223 Promotoren als am effizientesten in Hinsicht auf die Expression des gp91^{phox} Transgens, ROS Produktion und NETs Bildung in humanen differenzierten X-CGD hämatopoetischen Stammzellen identifiziert.*

Summary

Chronic granulomatous disease (CGD) is a group of primary immunodeficiencies caused by mutations in genes encoding phagocyte nicotinamide adenine dinucleotide phosphate (NADPH) oxidase subunits (gp91^{phox}, p22^{phox}, p47^{phox}, p67^{phox}, and p40^{phox}). CGD phagocytes do not produce reactive oxygen species (ROS), kill microbes poorly and consequently patients are susceptible to recurrent life-threatening bacterial and fungal infections. *Aspergillus spp.* infections, which cause pneumonia and disseminated disease, are the leading cause of death in CGD patients. Up to now it was unclear how neutrophils control *Aspergillus* species in healthy individuals.

We therefore set out to study the role of NADPH oxidase dependent neutrophil extracellular traps (NETs) in anti-Aspergillus defense in CGD. NETs are released after neutrophil activation and are composed of chromatin (DNA and histones) decorated with granular and cytoplasmic proteins. NETs bind and trap bacteria and fungi and expose antimicrobial molecules. Generation of NETs requires ROS produced by the NADPH oxidase; therefore CGD neutrophils cannot release NETs. *We showed for the first time that the microbicidal pathway through NETs is efficient against* A. nidulans *conidia and hyphae in vitro, and that restoration of NET formation was achieved by complementation of NADPH oxidase function by gene therapy (GT) in a patient with X-linked CGD.* This aided clearing severe invasive *A. nidulans* infection *in vivo*. *We demonstrated the critical role of NET-associate calprotectin for dose-dependent fungistatic of fungicidal anti-Aspergillus activity by Zn^{2+} sequestration.*

Gene therapy (GT) for CGD aims at functional correction of the NADPH oxidase subunit defects by insertion of a normal copy of the responsible gene into autologous patient derived hematopoietic stem cells (HSC) to cure the disease. γ-retroviral GT vectors have been extensively studied in international collaboration in recent phase I/II clinical GT for X-linked (gp91^{phox}-deficient) CGD in our Division of Immunology/Hematology/BMT at University Children's Hospital Zurich, proving feasibility of this approach. However, unexpected side effects like transgene silencing and oncogene transactivation impose improvements in vector efficacy and safety for future clinical GT trials.

We therefore developed and functionally studied next generation γ-retroviral self-inactivating (SIN) vectors as candidates for future clinical X-CGD GT trials. Among the myelospecific vectors tested in vitro, *we identified the human MRP8 and miRNA-223 promoters as most efficient in driving gp91^{phox} expression, leading to ROS production and NET formation in differentiated human X-CGD hematopoietic stem cells.*

INTRODUCTION

Chapter overview

The main subject of the described work is the phagocyte nicotinamide adenine dinucleotide phosphate (NADPH) oxidase deficient disorder chronic granulomatous disease (CGD), with regard to the *role of neutrophil extracellular trap (NET) formation* in the CGD disease process and the *correction of NADPH oxidase function by gene therapy for CGD*. This introductory chapter, therefore, focuses on CGD, the NADPH oxidase, the role of NETs in innate immunity against infection, and retroviral vectors as gene delivery system for human haematopoietic stem cell gene therapy.

Chronic granulomatous disease

CGD is a primary immunodeficiency that affects the oxidative mechanism of microbial killing of phagocytic cells. It was first described in the 1950s, termed fatal granulomatous disease of childhood [1-3]. The defect is due to absent or severely impaired activity of the NADPH oxidase of phagocytes (neutrophils, monocytes, macrophages and eosinophils), resulting in inadequate generation of reactive oxygen species (ROS), necessary for microbicidal activity within the phagosome, and lack of NETs [4-6]. Patients usually present early in life with recurrent and often life-threatening bacterial and fungal infections, and may develop chronic inflammatory granuloma in viscera and skin.

CGD occurs with an estimated frequency of 1 in 200'000 live births as a consequence of mutations in mainly one of five genes encoding the structural subunits of the NADPH oxidase complex (gp91^{phox}, p22^{phox}, p47^{phox}, p40^{phox}, and p67^{phox}) [7]. Approximately 65% of all cases result from mutations in the gp91^{phox} encoding *CYBB* gene (X-CGD) localized on the X-chromosome, autosomal-recessive forms are caused in 25% by *NCF1* gene mutations encoding p47^{phox}, while only 5% of cases are due to mutations in p67^{phox} encoding *NCF2* and p22^{phox} encoding *CYBA* [8] (these four forms of CGD are referred to as X91, A47, A67, and A22, respectively, with X for X-linked and A for autosomal [9]). Recently, one family with p40^{phox} encoding *NCF4* mutation has been described [10].

The diagnosis of CGD is usually made by assessing NADPH oxidase function via measurement of nitroblue tetrazolium (NBT) reduction, or dihydrorhodamine (DHR) oxidation. DHR is preferable because of its relative ease of use, its ability to distinguish X-linked from autosomal patterns of CGD by flow cytometry, and its sensitivity to very low numbers of functional neutrophils [11].

Clinical features of CGD

CGD is characterized by predisposition to severe, recurrent infection with bacteria and fungi [12]. CGD patients usually present within the first years of life with cervical or inguinal lymphadenitis, liver abscess, osteomyelitis, pneumonia, or skin abscesses [13] (Table 1).

Rarely CGD is diagnosed in only adulthood [14,15]. The microorganisms responsible for the majority of infections in CGD are *S. aureus*, Gram-negative enteric bacilli (including *Serratia marcescens*, *Salmonella* species, and *Burkholderia cepacia*) and *Aspergillus* species. Catalase-negative bacteria are rarely involved in CGD infection because of microbe-generated H_2O_2 in the phagosomes of neutrophils [13,15,16].

Table 1. Most common manifestations of CGD by organ system

Organ	Frequency	Most common manifestation of CGD	Unique features	Most common infecting organisms
Lungs	80%	Pneumonia	Contiguous spread to chest wall, associated osteomyelitis of ribs or vertebral bodies	*Aspergillus*, *Staphylococcus*
Lymph nodes	60%	Suppurative adenitis	Thick, enhancing septa; calcifications in chronic disease	*Staphylococcus*
Liver	25-50%	Hepatic abscess	Multiple small abscesses without an inciting event	*Staphylococcus aureus*
Muscoskeletal	25%	Osteomyelitis	Multifocal; occurs in ribs, vertebral bodies, small bones of hands and feet	*Serratia marcescens*, *Aspergillus*
Gastrointestinal	17%	Bowel wall thickening, mimics inflammatory bowel disease	Thickening of the wall of the gastric antrum	Granulomatous inflammation occurs in the absence of identifiable infection
Genitourinary	10-13%	Urinary tract infections	Inflammatory pseudotumor of the bladder	Has not been reported
Head and neck	12-15%	Sinusitis	Fungal sinusitis	Has not been reported
Central nervous system	4-5%	Encephalitis, brain abscess, meningitis	Abscesses have typical imaging appearance	*Candida* meningitis

Adapted from [17].

Patients with CGD also suffer from a variety of inflammatory conditions [13,18,19], including granuloma formation. Granulomas may be of microscopic or macroscopic size (up to several centimeters). Microscopic granulomas are typically part of a diffuse inflammatory process, such as colitis, while macroscopic granulomas usually cause a localized pathology through mechanical disturbance, such as gastric outlet obstruction. Granuloma formation can affect various organs, with a preference for hollow viscera, such as colon, stomach, and bladder. A number of observations argue in favor of a non-infectious origin of CGD granulomas, while the primary mechanism of the increased inflammatory response remains poorly understood [20-22].

Pneumonia and/or sepsis due to *Aspergillus* and *Burkholderia* are the most common causes of death in CGD patients [13]. X-linked CGD patients have been reported to have more severe clinical complications and higher mortality rates (5% vs 2% per year) than those with A47 CGD [13]. Female carriers of X91 CGD, with as few as 10% NADPH oxidase positive neutrophils, are asymptomatic [12].

Conventional treatment of CGD

The current treatment options for CGD are summarized in Table 2. Long-term antibiotic and antimycotic prophylaxis is the mainstay of treatment for CGD patients. Interferon-gamma (IFN-γ) has been shown to increase NADPH oxidase activity in some rare variants of X91 CGD with splice site mutations by improving splicing efficiency, resulting in generation of small amounts of functional $gp91^{phox}$. Donor granulocyte transfusions have been used for the treatment of life-threatening bacterial and fungal infections [23], limited by the risk of alloimmunization to HLA antigens which may complicate subsequent allogeneic haematopoietic stem cell transplantation (HSCT).

Table 2. CGD: main treatment modalities

Modality	Indication	Duration	Drug
Antibiotic prophylaxis	Bacterial infections	Lifelong	Trimethoprim-sulfamethoxazole
	Fungal infections	Lifelong	Itraconazole
Empiric antibiotic treatment	Gram⁻ infections	Until pathogen identification	Teicoplanin
	Gram⁻ infections	Until pathogen identification	Ciprofloxacin
	Fungal infections	Until pathogen identification	Voriconazole
Interferon-γ prophylaxis	Recurrent infections	Lifelong	Interferon-γ
White cell transfusion	Severe refractory infections	Until recovery of antibody formation	G-CSF stimulated leukocytes
Anti-inflammatory treatment	Obstructing granuloma	7-10 days → taper	Prednisolone
Stem cell transplantation	Recurrent serious manifestations	LAF-isolation 2 months, isolation at home 6-9 months	HLA identical marrow transplant

LAF, laminar air flow; G-CSF, granulocyte colony-stimulating factor; HLA, human leukocyte antigen. Adapted from [36].

HSCT is currently the only curative option for CGD patients, provided an HLA-identical donor is available [12,24-26]. The overall HSCT success rate is 81%, with an overall mortality of 15% [27]. Unfortunately, an HLA-identical donor is only available for 52% of CGD patients [28], and graft-versus-host disease (GvHD) and inflammatory flare-up at infectious sites are major risks associated with HSCT [27]. New studies with non-myeloablative reduced-intensity conditioning and HLA-identical donors provide encouraging results, reducing toxicity induced by myeloablative conditioning and reducing the risk of GvHD [29,30].

Gene therapy approaches for patients needing HSCT but lacking a compatible HSC donor have been developed, and will be discussed later on.

Neutrophils

Neutrophils are professional phagocytes indispensable for defense against intruding microorganisms. They are generated in large number in the bone marrow and circulate in blood for a few hours. If microorganisms have successfully overcome the physical barriers provided by the skin and mucus membranes and gained access to tissues, signals generated by microbes and resident macrophages at sites of infection activate the local endothelial cells, which capture bypassing neutrophils and guide them across the endothelial cell lining. Products generated by live microorganisms and by their interaction with soluble proteins that recognize microorganisms guide neutrophils toward the microbes, which are taken up by phagocytosis.

Neutrophils degrade microorganisms using a combination of oxidative and non-oxidative mechanisms (Fig. 1). Once phagocytosed, microorganisms are sequestered in the phagolysosome. Subsequent activation of the NADPH oxidase system generates large quantities of ROS. Release of ROS into the phagolysosome constitutes the oxidative arm of the microbicidal action of neutrophils. Traditionally, ROS were thought to be responsible for the direct killing of microorganisms. Intracellular sequestration of microorganisms also induces the fusion of neutrophil granules with the phagolysosome and the release of antimicrobial peptides and proteases from the granules into the phagolysosome. The action of these antimicrobial peptides and proteases constitutes the non-oxidative arm of the microbicidal action of neutrophils [31], acting by directly disrupting the microbial cell membrane [32,33].

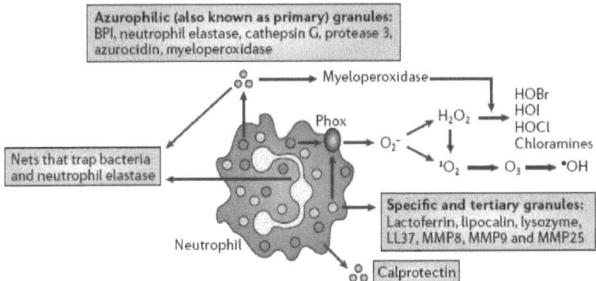

Figure 1. Neutrophils deliver multiple anti-microbial molecules from azurophilic granules (also known as primary granules), specific granules (also known as secondary granules) and tertiary granules, plasma and phagosomal membranes, the nucleus and the cytosol. BPI, bactericidal permeability increasing protein; H_2O_2, hydrogen peroxide; HOBr, hypobromous acid; HOCl, hypochlorous acid; HOI, hypoiodous acid; MMP, matrix metalloproteinase; $^1O_2$, singlet oxygen; O_2^-, superoxide; O_3, ozone; •OH, hydroxyl radical; phox, phagocyte oxidase. From [34].

Besides their direct antimicrobial activity, extracellularly released granule proteins contribute to microbial clearance by recruitment and activation of monocytes and macrophages [35] and by enhancing phagocytosis of pathogens by macrophages [36,37].

Neutrophil NADPH oxidase

The phagocyte NADPH oxidase consists of several proteins that are segregated between membrane and cytosol in resting cells (Fig. 2) [38-40]: flavocytochrome b558 is the central membrane-associated component of the NADPH oxidase and is composed of the glycosylated protein gp91^{phox} (phox for *ph*agocyte *ox*idase), non-covalently bound in a 1:1 stabilized complex to the p22^{phox} subunit, which is essential for maturation and stable expression of flavocytochrome b558. The cytosolic NADPH oxidase subunits are p40^{phox}, p47^{phox}, and p67^{phox}, which interact with each other (with a 1:1:1 stoichiometry) to form a complex, and the small G-proteins, Rac1 (in monocytes) or Rac2 (in neutrophils). The spatial separation of the NADPH oxidase components ensures that the enzyme is dormant in resting cells; in response to stimulation, the cytosolic components migrate almost instantly to the membrane where they assemble with flavocytochrome b558 to form the active enzyme, a process that is tightly regulated by protein-protein interactions and by phosphorylation [41,42].

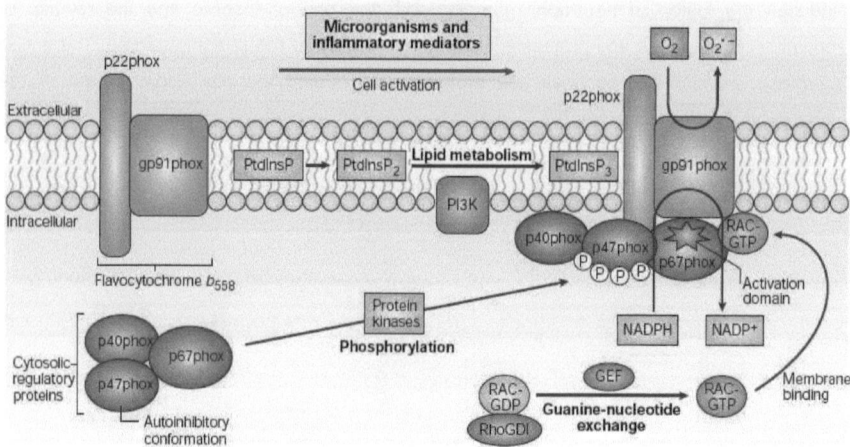

Figure 2. ROS generation by assembly of Phox proteins in phagocytes. Activation of the NADPH oxidase system occurs by at least three signaling triggers that result in the assembly of cytosolic regulatory proteins (p40^{phox}, p47^{phox} and p67^{phox}) with the membrane-associated catalytic subunits gp91^{phox} and p22^{phox} (flavocytochrome b558). These triggers involve protein kinases, lipid-metabolizing enzymes and nucleotide-exchange proteins that activate the GTPase RAC. Protein kinases catalyze phosphorylation of the autoinhibitory region of p47^{phox}, allowing its binding to p22^{phox}. Phosphatidylinositol 3-kinase (PI3K) produces 3-phosphorylated phosphatidylintositols (PtdInsP), providing lipids to which the p47^{phox} and p40^{phox} bind. Finally, activation of exchange factors (GEF) triggers GTP binding to RAC, resulting in conformational changes that promote dissociation from RhoGDI, membrane association and binding to p67^{phox}, helping to assemble the active complex. From [43].

The gp91^{phox} subunit is the electron transfer chain of the active NADPH oxidase and comprises binding sites for FAD, NADPH, and two hemes. Long thought to be specific to phagocytes, gp91^{phox} is now known to belong to a large family of proteins expressed in many different cell types and called NOX for NADPH oxidase, of which NOX2 is the phagocyte protein [44]. In resting cells, 60-70% of flavocytochrome b558 is located in the membranes of specific granules, 20-25% of flavocytochrome b558 is located in the membranes of gelatinase granules, while the remainder is found in plasma membranes and the membrane of secretory vesicles [45,46]. Properties of all NADPH oxidase components are summarized in table 3.

The NADPH oxidase subunits regulate their interaction with each other, with the membrane proteins and with lipids [41,42]. The structure of the cytosolic complex of resting cells is not entirely defined; however, p67^{phox} has been shown to associate tightly with p40^{phox}, and p47^{phox} has the ability to interact with both p40^{phox} and p67^{phox} [41,42]. In resting cells, Rac2, which is more abundant in human neutrophils than Rac1 (92% homologous to Rac2), is also present in the cytosol but is not part of the cytosolic complex. In its resting cytosolic GDP form, Rac2 is bound to its inhibitor, rho-GDI [47]. When neutrophils are activated and exocytosis of granules occurs, fusion of granule membranes with the plasma membrane increases the expression of flavocytochrome b558 [45,48], thought to be the central docking site for the 10–20% of the cytosolic NADPH oxidase components that translocate to the plasma membrane upon stimulation [47,49-51]. The activation is accompanied with extensive phosphorylation of p47[phox] on several serines located in the polybasic region of the carboxy-terminal portion of the protein [52], designated as the auto-inhibitory region. This results in the unmasking of cryptic SH3 domains, which can then bind the proline-rich region of p22[phox] [53]. Other interactions with gp91[phox]/NOX2 domains tether p47[phox] at the membrane. The p47[phox] subunit is thought to be responsible for transporting the cytosolic complex to the membrane during oxidase activation [49-51] and is considered as the organizer of the active NADPH oxidase complex. The p67[phox] subunit is also phosphorylated during activation [54], although to a lesser degree than p47[phox]. At the membrane, it binds to flavocytochrome b558 [55] and regulates its activity [56]. It also binds to Rac2 (or Rac1), which translocates to the plasma membrane independently and interacts with the p67[phox]/flavocytochrome b558 complex [41,42]. p40[phox] is weakly phosphorylated during activation [57]. It is not required for NADPH oxidase activation in a cell-free system but it has a positive stimulatory effect on enzyme activation *in vitro* [58]. On the contrary, p47[phox], p67[phox], and Rac, like flavocytochrome b558, are essential for optimal NADPH oxidase activation in a cell-free system and in intact neutrophils [38,39,41,42].

Table 3. Properties of the phagocyte respiratory burst oxidase components

	gp91[phox]	p22[phox]	p47[phox]	p67[phox]	p40[phox]	Rac2
Gene and locus	CYBB; Xp21.1	CYBA; 16q24	NCF-1; 7q11.23	NCF-2; 1q25	NCF-4; 22q13.1	Rac2; 22q13.1
Amino acids	570	195	390	526	339	192
Molecular weight:						
Predicted	65.338 kDa	20.959 kDa	44.684 kDa	59.735 kDa	39.039 kDa	21.429 kDa
By SDS-PAGE	90 kDa	22 kDa	47 kDa	67 kDa	40 kDa	22 kDa
Glycosylation	Yes	No	No	No	No	No
pI	9.26	10.1	9.58	6.12	7.28	7.87
Phosphorylation	No	Minor	Yes	Minor	Yes	Unknown
Location in PMN						
Resting	Specific granules and plasma membrane	Specific granules and plasma membrane	Cytosol	Cytosol	Cytosol	Mainly cytosol
Stimulated	Plasma membrane and phagosome	Plasma membrane and phagosome	Membrane	Membrane	Membrane	Membrane
Abundance pmol/10⁸ cells (cytosol conc.)	1.0-2.0	1.0-2.0	6.0 (2750 nM)	1.0 (460 nM)	1.0 (460 nM)	2.8 (1200nM)
Functional domains	C-terminus binds cytosolic components. Haem, FAD and NADPH binding regions	C-terminal proline-rich region	Phosphorylation sites, PX domain, 2 SH3 domains, proline-rich domain	Tetratricopeptide repeat, 2 SH3 domains, proline-rich domain	PX and SH3 domains, octicosapeptide repeat	GDP/GTP-binding; insert and effector regions, isoprenylation site

Adapted from [40].

Generation of reactive oxygen species

ROS production by activated NADPH oxidase begins with superoxide (O_2^-) formation by the following reaction [59]:

$$NADPH + 2O_2 \xrightarrow{\text{NADPH oxidase}} NADP^+ + H^+ + 2O_2^-$$

Two molecules of superoxide then react to generate hydrogen peroxide (H_2O_2) in a dismutation reaction, accelerated by the enzyme superoxide dismutase (SOD). In the presence of iron, superoxide and H_2O_2 react to generate hydroxyl radicals. In addition to superoxide, H_2O_2 and hydroxyl radicals, other ROS occur *in vivo*: in inflamed tissue, these include hypochlorous acid (HOCl), formed within neutrophils from H_2O_2 and chloride by the phagocyte enzyme myeloperoxidase (MPO); singlet oxygen, which might be formed from oxygen through the action of NADPH oxidase and MPO-catalysed oxidation of halide ions [60]; and ozone, which can be generated from singlet oxygen by antibody molecules (regardless of source of antigenic specificity) [61]. Formation and reactivity of ROS is schematically resumed in figure 3.

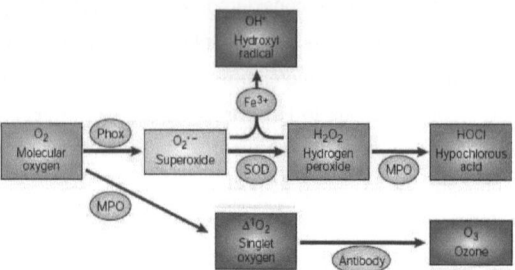

Figure 3. Superoxide generation. Superoxide is generated from various sources, which include the NADPH oxidase (phox). The color coding indicates the reactivity of individual molecules: from green (relatively unreactive) to red (high reactivity and non-specificity). From [43].

Neutrophil extracellular traps

In 2004, Arturo Zychlinsky et al described a new NADPH oxidase dependent antimicrobial mechanism of neutrophils, called neutrophil extracellular traps (NETs) [62], which extend neutrophil's antimicrobial action against bacteria [62], parasites [63], and fungi [4,64] beyond neutrophil death. Formation of NETs (Fig. 4) is an alternative to death by necrosis or apoptosis. During so called netosis, the nuclei swell and the chromatin is dissolved. Large strands of decondensed DNA are extruded from the cell, carrying along proteins from cytosol, granules, and histones [6,62,65]. 24 neutrophil proteins have been shown associated with NETs [66]: NET proteins are primarily the cationic (thus, DNA-binding) bactericidal proteins: histones, defensins, elastase, proteinase 3, heparin binding protein, cathepsin G, lactoferrin, and myeloperoxidase [66], but also the pattern recognition molecule Pentraxin 3 [67].

The mechanism of NET formation is not yet completely known. It has been shown that decondensation of chromatin is associated with citrullination of Histone H3 by conversions of histone arginine to citrulline residues by peptidylarginine deiminase 4 (PAD4) [68], an enzyme which is particularly rich in mature neutrophils [69]. Elastase released from azurophil granules degrades histones and synergizes with myeloperoxidase to drive chromatin decondensation [70]. NET formation is dependent on H_2O_2 generated by NADPH oxidase and further metabolized by myeloperoxidase (MPO). CGD neutrophils and MPO-deficient neutrophils, therefore, do not form NETs [4,6,71]. Moreover, the NADPH oxidase-activated Raf-MEK-ERK pathway is implicated in NET formation by inhibiting apoptosis to allow netosis [72].

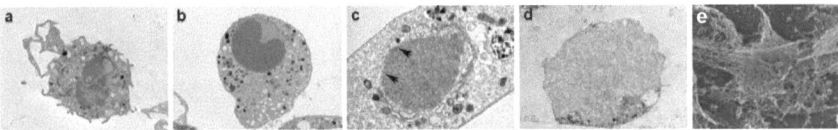

Figure 4. NET formation. Transmission (a-d) and scanning (e) electron microscopy of PMA-activated neutrophils. (a) Unstimulated neutrophils show a lobulated nucleus with clearly defined eu- and heterochromatin and numerous cytoplasmic vacuoles. (b) After 60 min of stimulation, nuclei are less lobulated. (c) After 120 min, most nuclei no longer show separation of eu- and heterochromatin, and in some cells the nuclear envelope starts to disintegrate into a chain of vesicles surrounding the DNA (arrows). (d) After 180 min of stimulation, most neutrophils are nearly entirely filled with decondensed chromatin, ready to be released. (e) Released NET-DNA after 240 min of stimulation. Adapted from [6].

The full contribution of NETs to overall antimicrobial activity versus the classical microbicidal activity of phagocytosing neutrophils *in vivo* has not yet been established. There is also evidence that NETs can contribute to pathological conditions: release of NETs in the blood stream has been reported in case of severe sepsis, ensnaring bacteria in the circulation but with the expense of injury to endothelium and tissues [73]. Moreover, the release of high levels of DNA and cellular content could play a role during development of autoimmune diseases like systemic lupus erythematosis (SLE). SLE patients have autoantibodies against DNA-associated proteins such as the ones found in NETs. A subset of patients has been identified with decreased activity of DNase-1, a NET-degrading protein found in plasma [74]. These patients had enhanced risk of nephritis. Furthermore, it was shown that SLE NETs participate to the chronicle activation of plasmacytoid dendritic cells (pDC) [75,76], a key point in SLE pathogenesis. It was also reported that NETs can induce thrombosis by stimulating platelets *in vitro*, and NET components were abundant in thrombi induced in a baboon model [77]. NETs were also found to damage activated human endothelial cells in culture [78], and seem to be involved in the pathogenesis of preeclampsia [79] and psoriasis [80]. All these observations point to a potential proinflammatory role of NETs.

Gene therapy

Gene therapy (GT) for primary immunodeficiency disease aims at functional correction of a genetic defect, to date by insertion of a normal and active copy of the responsible gene (transgene) into autologous patient hematopoietic stem cells (HSC) to cure the disease. Different strategies have been adopted to functionally correct the defective genes of interest. Replication-incompetent retroviral vectors are presently the most frequent studied tools for delivery of a therapeutic transgene into HSC. Retroviral transgene vectors enter the target cell, reverse transcribe and integrate the therapeutic gene into the host genome, from where it is expressed (Fig. 5).

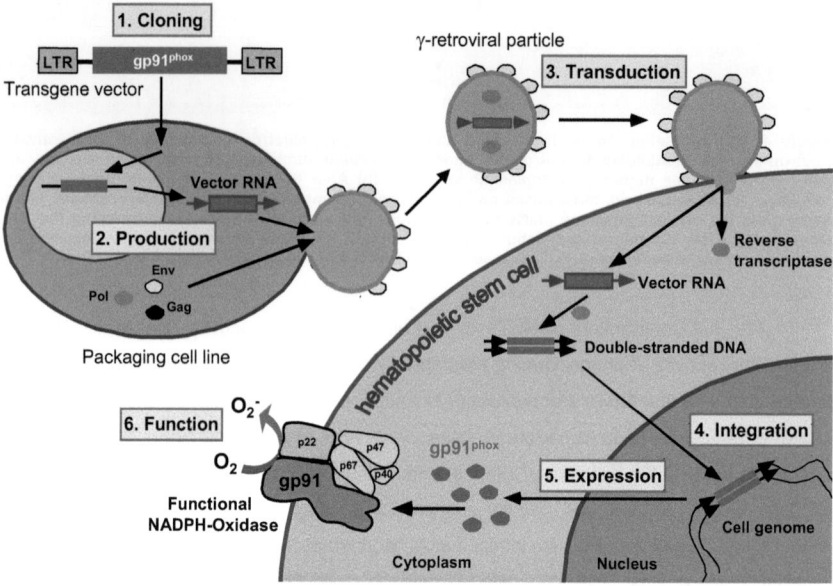

Figure 5. Schematic representation of a γ-retrovirally mediated X-CGD gene therapy.

To selectively target HSC with the retroviral vector and to avoid transduction of other tissues, standard GT uses G-CSF mobilized HSC isolated by apheresis from patient's peripheral blood. HSC are then retrovirally transduced *ex vivo*. At the end of the transduction period, corrected cells and the subset of cells that were not susceptible to transduction are reinfused into the patient and home to the bone marrow. Engraftment of the corrected cells and clinical success of gene therapy depend on: i) the transduction efficiency of the most primitive HSC [81]; ii) an eventual selective growth advantage of the corrected cells; iii) the chimerism of corrected and non-corrected (resident cells and transplanted non-transduced cells) engrafting cells; and iv) the persistence of transgene expression. The level of transgene expression is critical for functional correction. It depends on the promoter used and can later be hampered by methylation of the enhancer/promoter and by surrounding chromatin elements. The chimerism (percentage of transgene carrying cells) and the level of transgene expression are particularly critical for diseases such as CGD, where no selective growth advantage is conferred by the transgene.

Native retroviruses

Retroviridae are a family of single-stranded (ss) RNA viruses. The retroviral particle contains two copies of positive-strand RNA, which are complexed with nucleocapsid protein, together with the enzymes reverse transcriptase, integrase, and protease. A second protein shell, formed by capsid protein, encloses the nucleocapsid and delimits the viral core. Matrix proteins form a layer outside the core and interact with a cell-derived lipid envelope which incorporates viral envelope glycoproteins (env), responsible for the interaction with specific cellular receptors. Two units form these glycoproteins: the transmembrane, which anchors the protein into the lipid bilayer, and the surface, which binds to the cellular receptor.

Based on their genome organization, the *Retroviridae* are divided in simple and complex retroviruses. Examples are oncoretroviruses, such as murine leukemia virus (MLV), and lentiviruses, such as human immunodeficiency virus (HIV), respectively. Their genome is organized in the *gag*, *pro*, *pol*, and *env* genes (Fig. 6A). *Gag* (group-specific antigen) encodes the structural proteins, *pro* (protease) encodes the protease essential for gag cleavage during maturation, and *pol* (polymerase) gene encodes the enzymes that accompany the ssRNA. Of these, reverse transcriptase carries out reverse transcription of the viral RNA to DNA, integrase catalyses the integration of the proviral DNA into the host genome, and protease is involved in gag-pol cleavage and virion maturation. *Env* (envelope) encodes the viral envelope. In addition, complex retroviruses have accessory genes whose concerted action regulates viral gene expression, assembly, and replication. Moreover, the retroviral genome contains cis-acting sequences, such as two long terminal repeats (LTR), with elements required for gene expression. Other important components are the packaging signal (psi or ψ), required for the specific RNA packaging into newly formed virions, and the polypurine tract (PPT), which is the site of the initiation of positive-strand DNA synthesis during reverse transcription.

The retroviral life cycle starts with the binding of the viral envelope to its cellular receptor and fusion of the viral envelope with the cell membrane. Subsequently, the particle is uncoated and the viral core released into the cytoplasm. The ssRNA is reverse-transcribed into double-stranded (ds) DNA within the core, which is transported to the nucleus upon cell division for oncoretroviruses, or through active transport in the case of lentiviruses. The major advantage of lentiviruses over oncoretroviruses is their ability to transduce non-dividing cells. Once in the nucleus, the viral DNA is integrated into the host DNA (provirus), resulting in long-term expression of viral genes which are transcribed and spliced during the life of the infected cell. The full-length viral RNA as well as the RNA encoding all the viral proteins are transported to the cytoplasm, where they are translated. The unspliced full-length viral RNA is packaged and a viral particle assembled. Virion maturation occurs together with budding of the particle from the cell, as such resulting in new infectious virions.

Retroviral vectors as gene delivery system

The use of gene delivery vectors based on retroviruses was introduced in the early 1980s [82]. The most commonly used retroviral vectors are based on MLV (Fig. 6B). They can transfer 7 to 9 kb of foreign genetic information, which is sufficient for most therapeutic transgenes. The retroviral vectors used in GT applications are replication defective. In these vectors the genetic information elementary for virus production (*gag, pro, pol* and *env* genes) is deleted and is provided *in trans* during virus production by packaging cell lines. By changing virus envelope proteins the target cell specificity of the viral vector can be modified: for clinical HSC transduction, vectors pseudotyped with amphotropic MLV or gibbon ape leukemia virus (GALV) envelope have been used. HSC are heterogeneous with respect to their susceptibility to retroviral transduction, as transduction efficiency mainly depends on the type of retrovirus (gamma (γ)-, lenti-, e.g. MLV or HIV, respectively), the virus vector envelope (pseudotyping) and the cell cycle status of the HSC [83]. Unlike γ-retroviruses, which are dependent on cell division for integration, lentiviruses are able to transverse the intact nuclear membrane of a non-dividing HSC but are still more efficient at entering the nucleus during mitosis [84]. As HSC are quiescent at initiation of *ex vivo* culture, efficient transduction protocols with γ-retrovirus vectors require 3-4 days of *ex vivo* culture; whereas transduction protocols with lentiviral vectors can be as short as 1-2 days. In the last 15 years much progress has been achieved in the development of viral vectors derived from lentiviruses (such as HIV). A very important issue for lentiviral vectors has always been the biosafety with respect to the pathogenicity of the parental virus. The wild type HIV virus encodes the *vif, vpr, vpu* and *nef* genes. Their products are dispensable for virus production but are involved in HIV pathogenesis. In third generation lentiviral vectors these genes have been deleted, strongly improving the safety of these vectors [85-87].

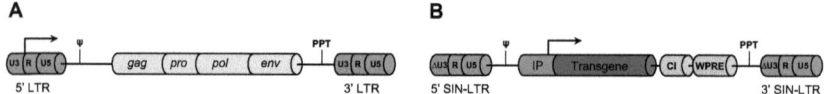

Figure 6. Retroviral vector development for GT. (A) Structure of a simple retroviral genome (e.g. MLV) containing coding sequences for *gag, pro, pol,* and *env* for replication. (B) Structure of a γ-retroviral self-inactivating (SIN) vector for GT. IP, internal (tissue-specific) promoter; CI, chromatin insulator; WPRE, woodchuck hepatitis virus post-transcriptional element.

First successful clinical gene therapy trials for primary immunodeficiency

The first successful clinical GT was reported by Fischer and Cavazzana-Calvo in 2000 [88] in two patients with the X-linked form of severe combined immune deficiency (SCID-X1) due to deficiency of the common gamma chain (γc) cytokine receptor. Overall, ten patients with SCID-X1 were treated, and achieved long-lasting immune reconstitution. The potentially curative outcome of gene therapy for SCID-X1 was later confirmed by Gaspar and Thrasher in Great Britain, who reported successful reconstitution of immunity in ten patients with SCID-X1 [89,90].

Bone marrow conditioning for GT was introduced in 2002 by Aiuti and colleagues who treated adenosine deiminase (ADA)-deficient SCID patients with low dose busulfan prior to infusion of transduced $CD34^+$ HSC, with the goal of improving the engraftment of gene corrected cells [91]. To date, data on thirty patients have been reported [92] indicating long-term multilineage engraftment of gene-corrected cells, associated with improvement in T-cell counts and proliferation to exogenous stimuli.

Both SCID-X1 and ADA-SCID GT trials lead to clear clinical benefits. Unfortunately, serious adverse events have been reported for five SCID-X1 patients, who developed leukemia due to transactivation events caused by the retroviral vector (insertional mutagenesis) [93-95]. No serious adverse events due to transactivation have been observed in the ADA-SCID trials so far [92].

Gene therapy for CGD

CGD is considered a suitable candidate for a GT approach, as all genes encoding for the subunits of the NADPH oxidase have been cloned. Moreover, data from variant forms of CGD and from healthy carriers of X91 CGD with as few as ≥ 10% normal neutrophils suggest that functional correction of a minor fraction of CGD neutrophils could be sufficient to cure the disease [8]. This observation together with a potential synergistic effect on antifungal activity of gene-corrected neutrophils and defective neutrophils in antifungal activity [96] have motivated the development of a GT protocol for CGD [97].

Early unsuccessful CGD gene therapy trials

Clinical GT trials for CGD were first initiated in the mid-1990s. The first of these were conducted by Malech and colleagues at the National Institutes of Health (USA): five patients with autosomal-recessive CGD ($p47^{phox}$ deficiency) were transfused with autologous $CD34^+$ HSC after genetic modification with a $p47^{phox}$-expressing γ-retroviral vector [98]. After transduction, HSC (0.1-4.7 x 10^6 cells/kg) were reinfused into the patients without bone marrow conditioning. Although the level of functionally corrected granulocytes after *in vitro* differentiation of transduced HSC was high (21-90%), the percentage of functionally corrected granulocytes circulating *in vivo* was low (0.004 to 0.05% of total peripheral blood granulocytes) and persisted at this level for up to 6 months after reinfusion. The same group initiated a similar trial for X-CGD in 1998. Several modifications were included in this protocol, including enhanced mobilization of $CD34^+$ HSC using Flt3-ligand (Flt3L) and granulocyte-macrophage colony-stimulating factor (GM-CSF), and retroviral transduction performed on 4 subsequent days resulting in an initial transduction efficiency of 48 to 89%.

Despite these modifications, the level of functionally corrected cells in peripheral blood still only ranged between 0.2 and 0.6% at 3-4 weeks after reinfusion and remained at this level for the next 4-6 months [97]. A third similar study was conducted by Dinauer and colleagues at Indiana University (USA). Autologous CD34$^+$ HSC from two adult X-CGD patients were transduced using a murine stem cell virus-based bicistronic γ-retroviral vector (MSCV-91Neo) containing the gp91^{phox} and the neomycin-resistance genes using a standard retronectin-based protocol. In this case, superoxide production was detected in both patients at up to 0.1% of peripheral blood neutrophils and persisted at this level for almost 9 months post-infusion (unpublished data, see.[99]).

One common denominator in these early clinical trials was the lack of bone marrow conditioning or myelosuppression that is conventionally used during allogeneic transplantation procedures [27]. Because gene-corrected CGD cells are not predicted to have a selective advantage over non-transduced cells (in contrast to SCID), engraftment of sufficient numbers of HSC to provide long-term correction is now considered to be possible only when bone marrow conditioning is performed or when a marker gene is co-expressed for *in vivo* selection.

Later successful CGD gene therapy trials

In 2006 bone marrow conditioning was successfully adopted on a study reported by Ott et al. and Stein et al, in which our Division of Immunology/Hematology/BMT at Zurich University was implied. CD34$^+$ cells obtained from two young adult X-CGD patients (age 25 and 26 years) were transduced using a γ-retroviral vector (SF71gp91^{phox}) containing a gp91^{phox} complementary DNA under the transcriptional control of the spleen focus-forming virus (SFFV) LTR [100,101] (the protocol used is summarized in figure 7).

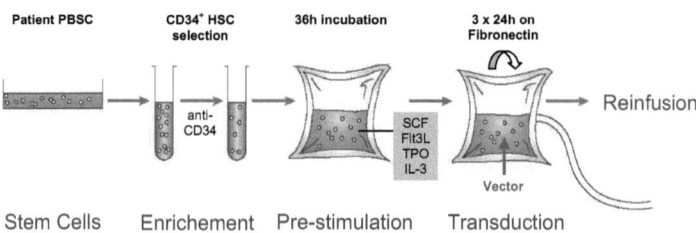

Figure 7. Zurich-Frankfurt X-CGD GT protocol. Isolated peripheral blood stem cells (PBSC) of CGD patients were first enriched by CD34$^+$ selection and then pre-stimulated during 36 hours with a cocktail of cytokines (SCF, Flt3L, TPO and IL-3) to induce cell division. CD34+ HSC were retrovirally transduced 3 times over 3 days and finally reinfused into the patients after low-dose Busulfan bone marrow conditioning.

Transduction efficiencies ranged between 40 and 45%, with 65-95% of the cells retaining expression of CD34 at the end of the 5-day transduction period. Transduced cells were reinfused at a dose between 9.0 and 11.3 x 10^6 CD34$^+$ cells/kg after reduced intensity conditioning with liposomal busulfan at a total dose of 8 mg/kg. Approximately 15% of peripheral blood neutrophils were found to express gp91^{phox} within the first 5 months after transplantation, this number increasing thereafter. The same vector and protocol were successfully used later by our Division to treat two children with X-CGD (manuscript in preparation). One of the children, an 8.5-year-old boy, suffered from therapy refractory *Aspergillus nidulans* lung infection [4]. Transduction efficiency reached 33% and the patient was transplanted with 18.2 x 10^6 CD34$^+$ total cells/kg after reduced intensity conditioning with busulfan intravenously (total dose 8.8 mg/kg). Gene marking in peripheral blood granulocytes was around 20% at day 20 after GT and remained stable until day 86, the last observation day reported in this study. Functionally corrected neutrophils were detected in the peripheral blood of this patient at levels between 30% (day 20) and 16% (day 86). The life-threatening, therapy refractory *A. nidulans* infection was cleared during the first six weeks after GT.

Risks of gene therapy using γ-retroviral vectors

Unfortunately, despite encouraging clinical results, the first GT trials for SCID-X1 also highlighted substantial risks of the procedure: considering all patients treated in the French and British trials, 5 patients out of 20 developed leukemia caused by the insertion of the therapeutic vector near LMO2, a known T-cell oncogene. Deregulation of LMO2 induced by the murine leukemia virus (MLV) enhancer was concluded to be the basis of the leukemic event [93-95,102].

In the CGD GT trial reported by Ott et al., the percentage of peripheral blood neutrophils expressing gp91^{phox} increased from 5 months after transplantation due to the insertional activation of growth-promoting genes, in particular PRDM16 and MDS1/EVI1 [100]. The overexpression of EVI1 lead to clonal dominance and ultimately monosomy 7 and myelodysplasia in both adult patients [101]. Moreover, a gradual loss of NADPH oxidase expressing neutrophils occurred 8 months after GT, caused by epigenetic inactivation (silencing) of the vector resulting in <5% superoxide-producing neutrophils (month 26 and 45, respectively). Gene-marked cells were still detectable, but CpG methylation within the promoter of the viral LTR resulted in reduced gp91^{phox} transgene expression. Enhancer elements of the viral LTR were not methylated and lead to oncogene transactivation.

Future of γ-retroviral gene therapy

The occurrence of unexpected severe adverse events in SCID-X1 and CGD GT trials has stimulated extensive preclinical research aimed at reducing the risk of insertional oncogenesis and vector silencing in future applications. The tendency of γ-retroviral vectors to integrate near transcription start sites of transcriptionally active genes [90,103-106] and the intrinsic promoter enhancing activity of retroviral LTRs are now known to be critical risk factors. To overcome the latter, self inactivating (SIN) [107] γ-retroviral vectors have been developed (Fig. 6B) for GT of SCID-X1 [108]. SIN-vectors lack the enhancer sequences of retroviral LTRs and are therefore less prone to activating promoters near integration sites [109,110]. These vectors require an internal promoter to drive transgene expression. Cell culture [111] and *in vivo* [112] assays showed that SIN-vectors with a strong viral internal enhancer/promoter have significantly lower transformation frequency by insertional mutagenesis compared to their LTR-driven counterparts [109]. Replacement of the strong viral promoter with weaker and tissue-specific cellular promoters further reduced transactivation activity [113]. The use of myeloid-specific promoters in CGD GT could target transgene expression to differentiated phagocytes and could at the same time avoid the use of strong enhancers potentially active in HSC. Addition of the ubiquitously-acting chromatin opening element (UCOE), which displays no enhancer function, has also been shown to have reduced transactivational activity compared to viral LTR promoters. Moreover, UCOE has been shown to resist silencing and to confer persistent transgene expression in a mouse study [114]. In addition, chromatin insulators such as the chicken β-globin hypersensitive site 4 have been used successfully to shield the surrounding chromatin from the vector and vice versa the vector from influences by the surrounding chromatin, resulting in less position-dependent silencing and in less position-effect variegation [115].

Objectives of the study

CGD patients are highly susceptible to severe recurrent *Aspergillus* infection. Despite the routine use of antifungal prophylaxis, life-threatening *Aspergillus* infections have remained a serious problem. *A. fumigatus* and *A. nidulans* are the most common *Aspergillus* species in CGD, with *A. nidulans* infection being associated to significantly higher mortality (50% versus 6%). Up to now it was unclear how neutrophils control *Aspergillus* hyphae, which are too large for phagocytosis, in healthy individuals. We hypothesized that NADPH oxidase-dependent NET formation might be involved in anti-*Aspergillus* defense, as NET formation is defective in CGD.

Objective 1: To determine whether gene therapy (GT) restores NET formation in CGD by complementation of NADPH oxidase function, and whether NETs have antimicrobial activity against *Aspergillus nidulans* (see chapter 1).

Objective 2: To determine the antifungal agent and mechanism responsible for reconstitution of *Aspergillus* growth inhibition within NETs after complementation of NADPH oxidase function by GT for CGD (see chapter 2).

To date allogeneic HSCT from an HLA identical donor is the only proven curative treatment for CGD, but is hampered by the availability of suitable HSC donors (only available for 52% of patients) and conditioning-regimen related toxicity. GT approaches to cure CGD have been clinically successful, but still face two major risks: oncogene transactivation and silencing of the transgene. Future GT vectors therefore have to balance sufficient transgene expression/efficiency with a minimum of transactivation activity.

Objective 3: To develop new GT (SIN) vectors with improved safety and efficiency profiles, for next generation phase I/II clinical CGD GT trials, and to develop a model to study NET formation after gene correction of CGD cells (see chapter 3).

References

1. Janeway C, Craig J, Davidson M, Downey W, Gitlin D, Sullivan J. Hypergammaglobulinemia associated with severe, recurrent and chronic non-specific infection. Am J Dis Child. 1954;88:388-392.
2. Berendes H, Bridges RA, Good RA. A fatal granulomatosus of childhood: the clinical study of a new syndrome. Minn Med. 1957;40:309-312.
3. Bridges RA, Berendes H, Good RA. A fatal granulomatous disease of childhood; the clinical, pathological, and laboratory features of a new syndrome. Am J Dis Child. 1959;97:387-408.
4. Bianchi M, Hakkim A, Brinkmann V, et al. Restoration of NET formation by gene therapy in CGD controls aspergillosis. Blood. 2009;114:2619-2622.
5. Bianchi M, Niemiec MJ, Siler U, Urban CF, Reichenbach J. Restoration of anti-Aspergillus defense by neutrophil extracellular traps in human chronic granulomatous disease after gene therapy is calprotectin-dependent. J Allergy Clin Immunol. 2011;127:1243-1252 e1247.
6. Fuchs TA, Abed U, Goosmann C, et al. Novel cell death program leads to neutrophil extracellular traps. J Cell Biol. 2007;176:231-241.
7. Seger RA. Chronic granulomatous disease: recent advances in pathophysiology and treatment. Neth J Med. 2010;68:334-340.
8. Holland SM. Chronic granulomatous disease. Clin Rev Allergy Immunol. 2010;38:3-10.
9. Heyworth PG, Cross AR, Curnutte JT. Chronic granulomatous disease. Curr Opin Immunol. 2003;15:578-584.
10. Matute JD, Arias AA, Wright NA, et al. A new genetic subgroup of chronic granulomatous disease with autosomal recessive mutations in p40 phox and selective defects in neutrophil NADPH oxidase activity. Blood. 2009;114:3309-3315.
11. Elloumi HZ, Holland SM. Diagnostic assays for chronic granulomatous disease and other neutrophil disorders. Methods Mol Biol. 2007;412:505-523.
12. Segal BH, Leto TL, Gallin JI, Malech HL, Holland SM. Genetic, biochemical, and clinical features of chronic granulomatous disease. Medicine (Baltimore). 2000;79:170-200.
13. Winkelstein JA, Marino MC, Johnston RB, Jr., et al. Chronic granulomatous disease. Report on a national registry of 368 patients. Medicine (Baltimore). 2000;79:155-169.
14. Cathebras P, Sauron C, Morel F, Stasia MJ. An unusual case of sarcoidosis. Lancet. 2001;358:294.

15. Martire B, Rondelli R, Soresina A, et al. Clinical features, long-term follow-up and outcome of a large cohort of patients with Chronic Granulomatous Disease: an Italian multicenter study. Clin Immunol. 2008;126:155-164.
16. Reichenbach J, Lopatin U, Mahlaoui N, et al. Actinomyces in chronic granulomatous disease: an emerging and unanticipated pathogen. Clin Infect Dis. 2009;49:1703-1710.
17. Towbin AJ, Chaves I. Chronic granulomatous disease. Pediatr Radiol. 2010;40:657-668; quiz 792-653.
18. Foster CB, Lehrnbecher T, Mol F, et al. Host defense molecule polymorphisms influence the risk for immune-mediated complications in chronic granulomatous disease. J Clin Invest. 1998;102:2146-2155.
19. Meissner F, Seger RA, Moshous D, Fischer A, Reichenbach J, Zychlinsky A. Inflammasome activation in NADPH oxidase defective mononuclear phagocytes from patients with chronic granulomatous disease. Blood;116:1570-1573.
20. Chin TW, Stiehm ER, Falloon J, Gallin JI. Corticosteroids in treatment of obstructive lesions of chronic granulomatous disease. J Pediatr. 1987;111:349-352.
21. Levine S, Smith VV, Malone M, Sebire NJ. Histopathological features of chronic granulomatous disease (CGD) in childhood. Histopathology. 2005;47:508-516.
22. Morgenstern DE, Gifford MA, Li LL, Doerschuk CM, Dinauer MC. Absence of respiratory burst in X-linked chronic granulomatous disease mice leads to abnormalities in both host defense and inflammatory response to Aspergillus fumigatus. J Exp Med. 1997;185:207-218.
23. von Planta M, Ozsahin H, Schroten H, Stauffer UG, Seger RA. Greater omentum flaps and granulocyte transfusions as combined therapy of liver abscess in chronic granulomatous disease. Eur J Pediatr Surg. 1997;7:234-236.
24. Seger RA. Modern management of chronic granulomatous disease. Br J Haematol. 2008;140:255-266.
25. Dinauer MC, Lekstrom-Himes JA, Dale DC. Inherited Neutrophil Disorders: Molecular Basis and New Therapies. Hematology Am Soc Hematol Educ Program. 2000:303-318.
26. Goldblatt D, Thrasher AJ. Chronic granulomatous disease. Clin Exp Immunol. 2000;122:1-9.
27. Seger RA, Gungor T, Belohradsky BH, et al. Treatment of chronic granulomatous disease with myeloablative conditioning and an unmodified hemopoietic allograft: a survey of the European experience, 1985-2000. Blood. 2002;100:4344-4350.
28. Tiercy JM, Villard J, Roosnek E. Selection of unrelated bone marrow donors by serology, molecular typing and cellular assays. Transpl Immunol. 2002;10:215-221.

29. Gungor T, Halter J, Klink A, et al. Successful low toxicity hematopoietic stem cell transplantation for high-risk adult chronic granulomatous disease patients. Transplantation. 2005;79:1596-1606.
30. Suzuki N, Hatakeyama N, Yamamoto M, et al. Treatment of McLeod phenotype chronic granulomatous disease with reduced-intensity conditioning and unrelated-donor umbilical cord blood transplantation. Int J Hematol. 2007;85:70-72.
31. Levy O. Antimicrobial proteins and peptides: anti-infective molecules of mammalian leukocytes. J Leukoc Biol. 2004;76:909-925.
32. Huang HW, Chen FY, Lee MT. Molecular mechanism of Peptide-induced pores in membranes. Phys Rev Lett. 2004;92:198304.
33. Brogden KA. Antimicrobial peptides: pore formers or metabolic inhibitors in bacteria? Nat Rev Microbiol. 2005;3:238-250.
34. Nathan C. Neutrophils and immunity: challenges and opportunities. Nat Rev Immunol. 2006;6:173-182.
35. Lai Y, Gallo RL. AMPed up immunity: how antimicrobial peptides have multiple roles in immune defense. Trends Immunol. 2009;30:131-141.
36. Heinzelmann M, Platz A, Flodgaard H, Miller FN. Heparin binding protein (CAP37) is an opsonin for Staphylococcus aureus and increases phagocytosis in monocytes. Inflammation. 1998;22:493-507.
37. Soehnlein O, Kai-Larsen Y, Frithiof R, et al. Neutrophil primary granule proteins HBP and HNP1-3 boost bacterial phagocytosis by human and murine macrophages. J Clin Invest. 2008;118:3491-3502.
38. Babior BM. NADPH oxidase: an update. Blood. 1999;93:1464-1476.
39. El-Benna J, Dang PM, Gougerot-Pocidalo MA, Elbim C. Phagocyte NADPH oxidase: a multicomponent enzyme essential for host defenses. Arch Immunol Ther Exp (Warsz). 2005;53:199-206.
40. Cross AR, Segal AW. The NADPH oxidase of professional phagocytes--prototype of the NOX electron transport chain systems. Biochim Biophys Acta. 2004;1657:1-22.
41. Quinn MT, Gauss KA. Structure and regulation of the neutrophil respiratory burst oxidase: comparison with nonphagocyte oxidases. J Leukoc Biol. 2004;76:760-781.
42. Groemping Y, Rittinger K. Activation and assembly of the NADPH oxidase: a structural perspective. Biochem J. 2005;386:401-416.
43. Lambeth JD. NOX enzymes and the biology of reactive oxygen. Nat Rev Immunol. 2004;4:181-189.
44. Bedard K, Krause KH. The NOX family of ROS-generating NADPH oxidases: physiology and pathophysiology. Physiol Rev. 2007;87:245-313.

45. Borregaard N, Heiple JM, Simons ER, Clark RA. Subcellular localization of the b-cytochrome component of the human neutrophil microbicidal oxidase: translocation during activation. J Cell Biol. 1983;97:52-61.
46. Kjeldsen L, Sengelov H, Lollike K, Nielsen MH, Borregaard N. Isolation and characterization of gelatinase granules from human neutrophils. Blood. 1994;83:1640-1649.
47. Abo A, Webb MR, Grogan A, Segal AW. Activation of NADPH oxidase involves the dissociation of p21rac from its inhibitory GDP/GTP exchange protein (rhoGDI) followed by its translocation to the plasma membrane. Biochem J. 1994;298 Pt 3:585-591.
48. Borregaard N, Kjeldsen L, Sengelov H, et al. Changes in subcellular localization and surface expression of L-selectin, alkaline phosphatase, and Mac-1 in human neutrophils during stimulation with inflammatory mediators. J Leukoc Biol. 1994;56:80-87.
49. Clark RA, Volpp BD, Leidal KG, Nauseef WM. Two cytosolic components of the human neutrophil respiratory burst oxidase translocate to the plasma membrane during cell activation. J Clin Invest. 1990;85:714-721.
50. el Benna J, Ruedi JM, Babior BM. Cytosolic guanine nucleotide-binding protein Rac2 operates in vivo as a component of the neutrophil respiratory burst oxidase. Transfer of Rac2 and the cytosolic oxidase components p47phox and p67phox to the submembranous actin cytoskeleton during oxidase activation. J Biol Chem. 1994;269:6729-6734.
51. Quinn MT, Evans T, Loetterle LR, Jesaitis AJ, Bokoch GM. Translocation of Rac correlates with NADPH oxidase activation. Evidence for equimolar translocation of oxidase components. J Biol Chem. 1993;268:20983-20987.
52. el Benna J, Faust LP, Babior BM. The phosphorylation of the respiratory burst oxidase component p47phox during neutrophil activation. Phosphorylation of sites recognized by protein kinase C and by proline-directed kinases. J Biol Chem. 1994;269:23431-23436.
53. Groemping Y, Lapouge K, Smerdon SJ, Rittinger K. Molecular basis of phosphorylation-induced activation of the NADPH oxidase. Cell. 2003;113:343-355.
54. Benna JE, Dang PM, Gaudry M, et al. Phosphorylation of the respiratory burst oxidase subunit p67(phox) during human neutrophil activation. Regulation by protein kinase C-dependent and independent pathways. J Biol Chem. 1997;272:17204-17208.

55. Dang PM, Cross AR, Babior BM. Assembly of the neutrophil respiratory burst oxidase: a direct interaction between p67PHOX and cytochrome b558. Proc Natl Acad Sci U S A. 2001;98:3001-3005.
56. Nisimoto Y, Motalebi S, Han CH, Lambeth JD. The p67(phox) activation domain regulates electron flow from NADPH to flavin in flavocytochrome b(558). J Biol Chem. 1999;274:22999-23005.
57. Bouin AP, Grandvaux N, Vignais PV, Fuchs A. p40(phox) is phosphorylated on threonine 154 and serine 315 during activation of the phagocyte NADPH oxidase. Implication of a protein kinase c-type kinase in the phosphorylation process. J Biol Chem. 1998;273:30097-30103.
58. Kuribayashi F, Nunoi H, Wakamatsu K, et al. The adaptor protein p40(phox) as a positive regulator of the superoxide-producing phagocyte oxidase. Embo J. 2002;21:6312-6320.
59. Robinson JM. Reactive oxygen species in phagocytic leukocytes. Histochem Cell Biol. 2008;130:281-297.
60. Kanofsky JR. Singlet oxygen production by biological systems. Chem Biol Interact. 1989;70:1-28.
61. Wentworth P, Jr., McDunn JE, Wentworth AD, et al. Evidence for antibody-catalyzed ozone formation in bacterial killing and inflammation. Science. 2002;298:2195-2199.
62. Brinkmann V, Reichard U, Goosmann C, et al. Neutrophil extracellular traps kill bacteria. Science. 2004;303:1532-1535.
63. Guimaraes-Costa AB, Nascimento MT, Froment GS, et al. Leishmania amazonensis promastigotes induce and are killed by neutrophil extracellular traps. Proc Natl Acad Sci U S A. 2009;106:6748-6753.
64. Urban CF, Reichard U, Brinkmann V, Zychlinsky A. Neutrophil extracellular traps capture and kill Candida albicans yeast and hyphal forms. Cell Microbiol. 2006;8:668-676.
65. Brinkmann V, Zychlinsky A. Beneficial suicide: why neutrophils die to make NETs. Nat Rev Microbiol. 2007;5:577-582.
66. Urban CF, Ermert D, Schmid M, et al. Neutrophil extracellular traps contain calprotectin, a cytosolic protein complex involved in host defense against Candida albicans. PLoS Pathog. 2009;5:e1000639.
67. Jaillon S, Peri G, Delneste Y, et al. The humoral pattern recognition receptor PTX3 is stored in neutrophil granules and localizes in extracellular traps. J Exp Med. 2007;204:793-804.
68. Wang Y, Wysocka J, Sayegh J, et al. Human PAD4 regulates histone arginine methylation levels via demethylimination. Science. 2004;306:279-283.

69. Nakashima K, Hagiwara T, Yamada M. Nuclear localization of peptidylarginine deiminase V and histone deimination in granulocytes. J Biol Chem. 2002;277:49562-49568.
70. Papayannopoulos V, Metzler KD, Hakkim A, Zychlinsky A. Neutrophil elastase and myeloperoxidase regulate the formation of neutrophil extracellular traps. J Cell Biol. 2010;191:677-691.
71. Metzler KD, Fuchs TA, Nauseef WM, et al. Myeloperoxidase is required for neutrophil extracellular trap formation: implications for innate immunity. Blood. 2010.
72. Hakkim A, Fuchs TA, Martinez NE, et al. Activation of the Raf-MEK-ERK pathway is required for neutrophil extracellular trap formation. Nat Chem Biol. 2010.
73. Clark SR, Ma AC, Tavener SA, et al. Platelet TLR4 activates neutrophil extracellular traps to ensnare bacteria in septic blood. Nat Med. 2007;13:463-469.
74. Hakkim A, Furnrohr BG, Amann K, et al. Impairment of neutrophil extracellular trap degradation is associated with lupus nephritis. Proc Natl Acad Sci U S A. 2010;107:9813-9818.
75. Lande R, Ganguly D, Facchinetti V, et al. Neutrophils activate plasmacytoid dendritic cells by releasing self-DNA-peptide complexes in systemic lupus erythematosus. Sci Transl Med. 2011;3:73ra19.
76. Villanueva E, Yalavarthi S, Berthier CC, et al. Netting Neutrophils Induce Endothelial Damage, Infiltrate Tissues, and Expose Immunostimulatory Molecules in Systemic Lupus Erythematosus. J Immunol. 2011.
77. Fuchs TA, Brill A, Duerschmied D, et al. Extracellular DNA traps promote thrombosis. Proc Natl Acad Sci U S A. 2010;107:15880-15885.
78. Gupta AK, Joshi MB, Philippova M, et al. Activated endothelial cells induce neutrophil extracellular traps and are susceptible to NETosis-mediated cell death. FEBS Lett. 2010;584:3193-3197.
79. Gupta AK, Hasler P, Holzgreve W, Gebhardt S, Hahn S. Induction of neutrophil extracellular DNA lattices by placental microparticles and IL-8 and their presence in preeclampsia. Hum Immunol. 2005;66:1146-1154.
80. Lin AM, Rubin CJ, Khandpur R, et al. Mast Cells and Neutrophils Release IL-17 through Extracellular Trap Formation in Psoriasis. J Immunol. 2011.
81. Brenner S, Ryser MF, Whiting-Theobald NL, Gentsch M, Linton GF, Malech HL. The late dividing population of gamma-retroviral vector transduced human mobilized peripheral blood progenitor cells contributes most to gene-marked cell engraftment in nonobese diabetic/severe combined immunodeficient mice. Stem Cells. 2007;25:1807-1813.

82. Mann R, Mulligan RC, Baltimore D. Construction of a retrovirus packaging mutant and its use to produce helper-free defective retrovirus. Cell. 1983;33:153-159.
83. Brenner S, Malech HL. Current developments in the design of onco-retrovirus and lentivirus vector systems for hematopoietic cell gene therapy. Biochim Biophys Acta. 2003;1640:1-24.
84. Roe T, Reynolds TC, Yu G, Brown PO. Integration of murine leukemia virus DNA depends on mitosis. Embo J. 1993;12:2099-2108.
85. Dull T, Zufferey R, Kelly M, et al. A third-generation lentivirus vector with a conditional packaging system. J Virol. 1998;72:8463-8471.
86. Naldini L, Blomer U, Gallay P, et al. In vivo gene delivery and stable transduction of nondividing cells by a lentiviral vector. Science. 1996;272:263-267.
87. Zufferey R, Nagy D, Mandel RJ, Naldini L, Trono D. Multiply attenuated lentiviral vector achieves efficient gene delivery in vivo. Nat Biotechnol. 1997;15:871-875.
88. Cavazzana-Calvo M, Hacein-Bey S, de Saint Basile G, et al. Gene therapy of human severe combined immunodeficiency (SCID)-X1 disease. Science. 2000;288:669-672.
89. Gaspar HB, Parsley KL, Howe S, et al. Gene therapy of X-linked severe combined immunodeficiency by use of a pseudotyped gammaretroviral vector. Lancet. 2004;364:2181-2187.
90. Schwarzwaelder K, Howe SJ, Schmidt M, et al. Gammaretrovirus-mediated correction of SCID-X1 is associated with skewed vector integration site distribution in vivo. J Clin Invest. 2007;117:2241-2249.
91. Aiuti A, Slavin S, Aker M, et al. Correction of ADA-SCID by stem cell gene therapy combined with nonmyeloablative conditioning. Science. 2002;296:2410-2413.
92. Aiuti A, Roncarolo MG. Ten years of gene therapy for primary immune deficiencies. Hematology Am Soc Hematol Educ Program. 2009:682-689.
93. Hacein-Bey-Abina S, Garrigue A, Wang GP, et al. Insertional oncogenesis in 4 patients after retrovirus-mediated gene therapy of SCID-X1. J Clin Invest. 2008;118:3132-3142.
94. Hacein-Bey-Abina S, von Kalle C, Schmidt M, et al. A serious adverse event after successful gene therapy for X-linked severe combined immunodeficiency. N Engl J Med. 2003;348:255-256.
95. Hacein-Bey-Abina S, Von Kalle C, Schmidt M, et al. LMO2-associated clonal T cell proliferation in two patients after gene therapy for SCID-X1. Science. 2003;302:415-419.
96. Rex JH, Bennett JE, Gallin JI, Malech HL, Melnick DA. Normal and deficient neutrophils can cooperate to damage Aspergillus fumigatus hyphae. J Infect Dis. 1990;162:523-528.

97. Grez M, Reichenbach J, Schwable J, Seger R, Dinauer MC, Thrasher AJ. Gene therapy of chronic granulomatous disease: the engraftment dilemma. Mol Ther. 2011;19:28-35.
98. Malech HL, Maples PB, Whiting-Theobald N, et al. Prolonged production of NADPH oxidase-corrected granulocytes after gene therapy of chronic granulomatous disease. Proc Natl Acad Sci U S A. 1997;94:12133-12138.
99. Barese CN, Goebel WS, Dinauer MC. Gene therapy for chronic granulomatous disease. Expert Opin Biol Ther. 2004;4:1423-1434.
100. Ott MG, Schmidt M, Schwarzwaelder K, et al. Correction of X-linked chronic granulomatous disease by gene therapy, augmented by insertional activation of MDS1-EVI1, PRDM16 or SETBP1. Nat Med. 2006;12:401-409.
101. Stein S, Ott MG, Schultze-Strasser S, et al. Genomic instability and myelodysplasia with monosomy 7 consequent to EVI1 activation after gene therapy for chronic granulomatous disease. Nat Med. 2010;16:198-204.
102. Qasim W, Gaspar HB, Thrasher AJ. Update on clinical gene therapy in childhood. Arch Dis Child. 2007;92:1028-1031.
103. Deichmann A, Hacein-Bey-Abina S, Schmidt M, et al. Vector integration is nonrandom and clustered and influences the fate of lymphopoiesis in SCID-X1 gene therapy. J Clin Invest. 2007;117:2225-2232.
104. Mitchell RS, Beitzel BF, Schroder AR, et al. Retroviral DNA integration: ASLV, HIV, and MLV show distinct target site preferences. PLoS Biol. 2004;2:E234.
105. Wu X, Li Y, Crise B, Burgess SM. Transcription start regions in the human genome are favored targets for MLV integration. Science. 2003;300:1749-1751.
106. Hematti P, Hong BK, Ferguson C, et al. Distinct genomic integration of MLV and SIV vectors in primate hematopoietic stem and progenitor cells. PLoS Biol. 2004;2:e423.
107. Yu SF, von Ruden T, Kantoff PW, et al. Self-inactivating retroviral vectors designed for transfer of whole genes into mammalian cells. Proc Natl Acad Sci U S A. 1986;83:3194-3198.
108. Thornhill SI, Schambach A, Howe SJ, et al. Self-inactivating gammaretroviral vectors for gene therapy of X-linked severe combined immunodeficiency. Mol Ther. 2008;16:590-598.
109. Modlich U, Bohne J, Schmidt M, et al. Cell-culture assays reveal the importance of retroviral vector design for insertional genotoxicity. Blood. 2006;108:2545-2553.
110. Schambach A, Galla M, Maetzig T, Loew R, Baum C. Improving transcriptional termination of self-inactivating gamma-retroviral and lentiviral vectors. Mol Ther. 2007;15:1167-1173.

111. Du Y, Jenkins NA, Copeland NG. Insertional mutagenesis identifies genes that promote the immortalization of primary bone marrow progenitor cells. Blood. 2005;106:3932-3939.
112. Modlich U, Kustikova OS, Schmidt M, et al. Leukemias following retroviral transfer of multidrug resistance 1 (MDR1) are driven by combinatorial insertional mutagenesis. Blood. 2005;105:4235-4246.
113. Zychlinski D, Schambach A, Modlich U, et al. Physiological promoters reduce the genotoxic risk of integrating gene vectors. Mol Ther. 2008;16:718-725.
114. Zhang F, Thornhill SI, Howe SJ, et al. Lentiviral vectors containing an enhancer-less ubiquitously acting chromatin opening element (UCOE) provide highly reproducible and stable transgene expression in hematopoietic cells. Blood. 2007;110:1448-1457.
115. Aker M, Tubb J, Groth AC, et al. Extended core sequences from the cHS4 insulator are necessary for protecting retroviral vectors from silencing position effects. Hum Gene Ther. 2007;18:333-343.

CHAPTER 1

Restoration of NET Formation by Gene Therapy in CGD controls Aspergillosis

Matteo Bianchi[1*], Abdul Hakkim[2*], Volker Brinkmann[3], Ulrich Siler[1], Reinhard A. Seger[1], Arturo Zychlinsky[2+], Janine Reichenbach[1+]

*, + These authors contributed equally to the work

[1] Division of Immunology/Hematology/BMT, University Children's Hospital Zurich, Switzerland
[2] Department of Cellular Microbiology and [3] Microscopy Core Facility, Max Planck Institute for Infection Biology, Berlin, Germany

Corresponding Authors:
Janine Reichenbach
Division of Immunology/Hematology/BMT
University Children's Hospital Zurich
Steinwiesstrasse 75, 8032 Zurich, Switzerland
Phone: +41 44 266 7311, Fax: +41 44 266 7914
e-mail: janine.reichenbach@kispi.uzh.ch

Arturo Zychlinsky
Department of Cellular Microbiology
Max Planck Institute for Infection Biology
Charitéplatz 1, Berlin 10117, Germany.
Phone: +49 30 28460 300, Fax: +49 30 28460 301
e-mail: zychlinsky@mpiib-berlin.mpg.de

Published in: *Blood* 2009;114:2619-22

Own contribution: M.B. performed the experiments (except Fig.1A and C), analyzed data, prepared all figures and contributed to the writing of the manuscript.

Abstract

Chronic granulomatous disease (CGD) patients have impaired nicotinamide adenine dinucleotide phosphate (NADPH) oxidase function, resulting in poor antimicrobial activity of neutrophils including the inability to generate Neutrophil Extracellular Traps (NETs). Invasive aspergillosis is the leading cause of death in patients with CGD; it is unclear how neutrophils control *Aspergillus* species in healthy individuals. The aim of this study was to determine whether gene therapy (GT) restores NET formation in CGD by complementation of NADPH oxidase function, and whether NETs have antimicrobial activity against *Aspergillus nidulans*. Here we show that reconstitution of NET formation by GT in a patient with CGD restores neutrophil elimination of *Aspergillus nidulans* conidia and hyphae and is associated with rapid cure of pre-existing therapy refractory invasive pulmonary aspergillosis, underlining the role of functional NADPH oxidase in NET formation and antifungal activity.

Introduction

Activated neutrophils kill microbes intracellularly following phagocytosis and by extracellular mechanisms including Neutrophil Extracellular Traps (NETs), which are composed of chromatin decorated with granular proteins [1]. NETs bind bacteria [1] and fungi [2] and expose antimicrobial molecules. Generation of NETs requires reactive oxygen species (ROS) produced by the NADPH oxidase [3].

Chronic granulomatous disease (CGD) is caused by mutations in genes encoding NADPH oxidase subunits. CGD patients do not produce ROS, kill microbes poorly and are susceptible to recurrent life-threatening infections [4]. *Aspergillus spp.* infections cause pneumonia and disseminated disease and are the leading cause of death in these patients [4-6].

It is unclear how *Aspergillus* infections are controlled in healthy individuals [7-13]. In CGD patients, these infections are frequently refractory to antifungal therapy, treatment with interferon-γ, or granulocyte transfusions [5]. Here we show that the recently discovered NADPH oxidase dependent microbicidal pathway through NETs [1-3] is efficient against *A. nidulans* conidia and hyphae *in vitro*, and that restoration of NET formation by GT of X-CGD aided clearing severe invasive *A. nidulans* infection *in vivo*.

Methods

Gene therapy

We treated an 8.5-year-old boy with X-linked gp91^{phox}-deficient CGD and therapy refractory *A. nidulans* lung infection with a monocistronic long terminal repeat-driven gammaretroviral SF71gp91^{phox} vector (see supplemental data). The protocol for the patient's treatment was approved by the ethics review board of the University Children's Hospital Zurich and the

Swiss Expert Committee for Bio-Safety, after written informed consent from his parents in accordance with the Declaration of Helsinki. For follow-up monitoring, gp91^{phox} expression was measured by fluorescence-activated cell sorter (FACS) on peripheral neutrophils after 30 min staining at room temperature with 10 µg/ml gp91^{phox}-fluorescin isothiocyanate (FITC) antibody (Anti-Flavocytochrome b558, clone 7D5, MBL). NADPH oxidase activity was measured by standard dihydrorhodamine (DHR) and nitroblue tetrazolium (NBT) tests (supplemental materials). Bone marrow colony assays and determination of proviral gp91^{phox} sequences in genomic DNA were performed as described [14].

NET induction

NET formation was visualized as described (supplemental materials) and quantified after stimulation of 5x10^4 neutrophils for 3h with 40nM phorbol 12-myristate 13-acetate (PMA) and staining the NET-DNA with 1 µM Sytox green (Invitrogen) in a black 96-well plate (BD Biosciences). The plates were read in a fluorescence microplate reader (Victor^3, PerkinElmer Life and Analytical Sciences) with a filter setting of 485/535nm (excitation/emission).

NET antifungal activity

The *A. nidulans* strain used was isolated from bronchoalveolar lavage fluid of the patient; conidia were grown and collected as described [7]. Neutrophils after GT were stained with gp91^{phox}-FITC antibody and sorted by FACS (FACSAria, BD Biosciences) into gp91^{phox}-negative (gp91^{phox-}) and -positive (gp91^{phox+}) populations. 10^5 neutrophils were activated with PMA (40nM) at 37°C for 4h in a 96-well plate, then infected with conidia (multiplicity of infection *A. nidulans*/neutrophils 0.5) ± prior digestion of NETs with 10 U/ml of Micrococcal Nuclease (MNase, Worthington Biochemical) for 30 min. Afterwards the plates were centrifuged 5 min at 400 g and incubated for 16h at 37°C, allowing germination and hyphal outgrowth. Alternatively, 5x10^4 conidia were incubated for 12 h at 37°C to allow hyphal outgrowth; then 10^5 neutrophils were added, centrifuged 5 min at 400 g and incubated for 2 and 5 h with PMA (40nM) to induce NET formation, ± 10 U/ml MNase. Fungal growth was quantified with XTT (Invitrogen) as described [15].

Results and discussion

In healthy subjects NETs might be essential to eliminate fungi since hyphae are too large to be phagocytosed [7,9,12,13,16-21]. CGD patients are unable to make NETs [3]. Indeed, neutrophils of our X-CGD patient could only make NETs and inhibit growth of *A. nidulans* after genetic complementation by GT. Neutrophils expressing functional gp91^{phox} increased from 0% to 26-29% 6 weeks after GT (Figures 1A-B), then decreased and leveled around 16% for up to 3 months. The *A. nidulans* infection completely cleared 6 weeks after GT (Figure 1C), correlating with the rise in neutrophils with NADPH oxidase activity.

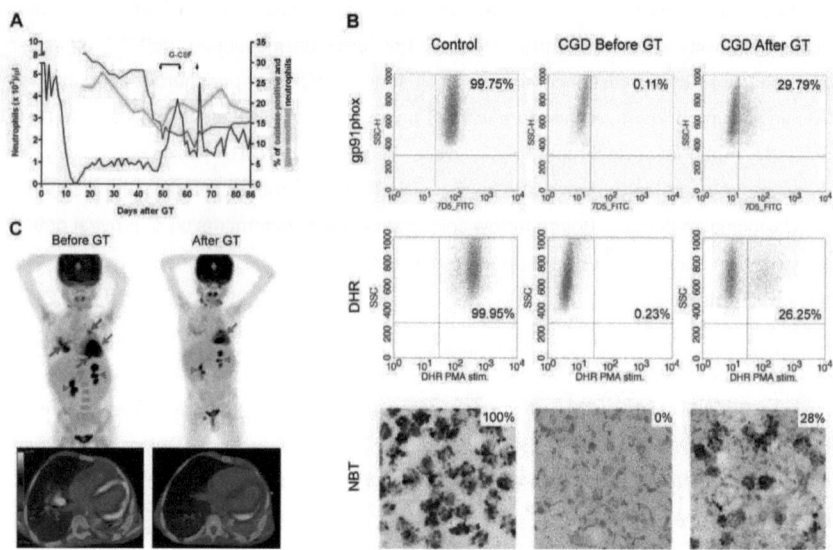

Figure 1. Restoration of NADPH oxidase function. (A) Haematopoietic reconstitution and gene marking after GT. Absolute neutrophil counts (left y-axis), quantification of gene-modified cells in peripheral neutrophils by Q-PCR and quantification of neutrophils with NADPH oxidase activity by DHR test (right y-axis) are shown. When the percentage of transduced neutrophils decreased, G-CSF (5 µg/KG/day subcutaneously) was administered on days + 49 to + 57 and on day + 64. (B) Reconstitution of NADPH oxidase activity. Before and 6 weeks after GT gp91^{phox} protein expression was measured by FACS analysis after 30 min staining with 10 µg/ml gp91^{phox}-FITC antibody. Superoxide production was assessed by oxidation of DHR upon stimulation with PMA and by reduction of NBT to formazan (dark precipitate) after stimulation with opsonized zymosan (OPZ). The thresholds were determined using unstained (FACS) or unstimulated (DHR) cells for each experiment. (C) PET-CT scan. Before GT, PET-CT scan showed several active infectious foci with fluorine-18-fluoro-2-deoxy-D-glucose (FDG) uptake in both lungs of the patient (red arrows); infection cleared 6 weeks after administration of gene corrected cells. In green, physiologic FDG uptake in heart (arrow), kidneys (arrowheads), bladder (asterisk) and brain (diamond) are indicated for reference.

The patient's neutrophils did not make NETs before GT as analyzed by fluorescence (not shown), immunofluorescence and scanning electron microscopy (Figures 2A-B and D-E) [3]. After GT, the patient's neutrophils made NETs (Figures 2C,F), the percentage of cells releasing NET-DNA (28%; Figure 2G) correlating with the level of oxidase chimerism (Figure 1B). Activation of sorted neutrophils showed that reconstitution with functional NADPH oxidase allowed corrected CGD neutrophils to make NETs (Figure 2H).

To test whether efficient eradication of the patient's infection was due to the recovered ability to make NETs, neutrophils were infected with the *A. nidulans* strain isolated from the patient. About 80% of conidia germination (Figure 2I) and 45% of hyphal growth (Figure 2J) were inhibited by CGD gp91^{phox+} neutrophils, comparable to the antimicrobial activity of control neutrophils. CGD gp91^{phox-} neutrophils were inefficient in controlling fungal growth (Figure 2I-J, supplemental Figure 1). When NETs were dismantled with MNase before infection, fungicidal activity was abrogated to that of CGD gp91^{phox-} neutrophils.

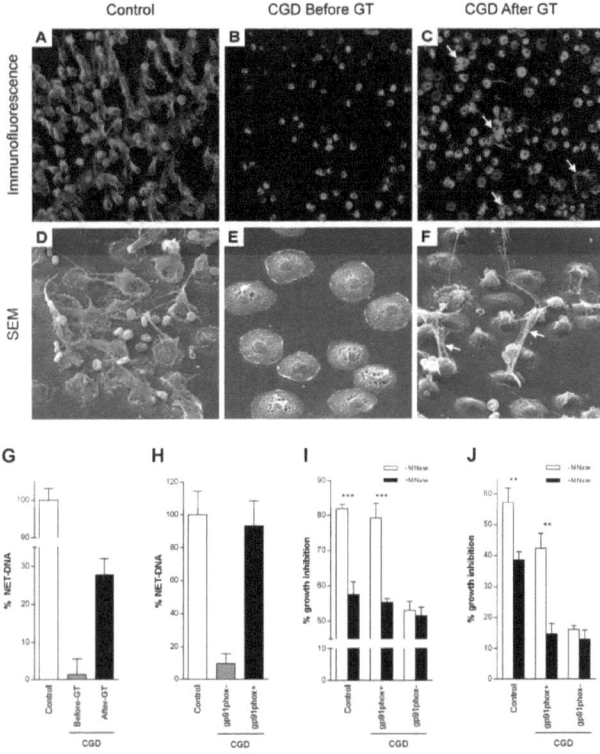

Figure 2. NET formation and inhibition of *A. nidulans* growth. Control (A,D), but not CGD (B,E) neutrophils made NETs upon 3h PMA stimulation. For immunofluorescence, NETs were stained with an antibody that recognizes neutrophil elastase (green; A-C). NETs were clearly visible also by Scanning Electron Microscopy (SEM; D-F). Neutrophils isolated from the CGD patient before GT could be activated since they flattened out (E), but did not make NETs. The ability to form NETs was partially restored by GT 6 weeks after GT (C,F white arrows). (G) Quantification of NET-DNA released after 3 h PMA stimulation of control neutrophils, CGD neutrophils before and 6 weeks after GT or (H) after stimulation of CGD gp91^{phox+} and CGD gp91^{phox-} FACS sorted neutrophils. CGD gp91^{phox+} neutrophils showed normal NET formation, while CGD gp91^{phox-} neutrophils showed only residual NET formation. FACS-sorting efficiency was 90 – 92% for CGD gp91^{phox-} and 95 – 96% for CGD gp91^{phox+} cells. (I, J) NET inhibition of *A. nidulans* conidia and hyphae. (I) Conidia were plated on FACS-sorted neutrophils pre-stimulated with PMA ± MNase i.e. when NET formation was complete, cells were dead and therefore incapable of phagocytosis. Hyphal outgrowth was measured after 16 h. (J) Hyphae were co-incubated with FACS-sorted neutrophils and PMA ± MNase and hyphal viability was assessed after 5h. The data shown in panels G to J are presented as mean ± SD of a representative triplicate experiment. Inhibition of fungal growth is expressed as percentage of control values (*A. nidulans* conidia or hyphae incubated in media). The differences between -MNase and +MNase were significant (for control and CGD gp91^{phox+} cells) by Student's *t*-test (**$P < .01$; ***$P < .001$).

In the absence of NETs control of hyphal growth was independent of NADPH oxidase activity: when neutrophils were infected after 2h PMA stimulation, before NETs had been made, CGD gp91^{phox+}, CGD gp91^{phox-} and neutrophils from healthy donor controlled growth of *A. nidulans* with similar modest efficiency (Figure S2) in a NET-independent fashion, since antimicrobial activity was not affected by MNase. This limited NET-independent antimicrobial activity, presumably by conidia phagocytosis, degranulation, or unknown mechanisms, suggests an NADPH oxidase-independent antifungal mechanism, clinically ineffective before GT. These data propose that the patient's clearance of fungal infection after GT was controlled by NETs. Definitive *in vivo* proof is obviously technically impossible.

Alveolar macrophages probably constitute the first line of defense to conidia that escape mucociliary clearance in healthy individuals [13,16]. Whether reconstituted NADPH oxidase function in alveolar macrophages also contributed to microbial killing in the patient presented is difficult to assess. In neutrophils, however, conidia resist intracellular killing due to their relative tolerance to ROS [7,9,12,19-22]. Our results suggest that conidia are killed mainly extracellularly rather than following phagocytosis. Both conidia and hyphae get ensnared by neutrophils and likely killed within NETs by concentrated antimicrobials. Cooperation of gp91^{phox-} and gp91^{phox+} neutrophils in NET antifungal activity is unlikely, since we showed that gp91^{phox-} neutrophils do not make NETs when co-incubated with gp91^{phox+} neutrophils (unsorted cells in Figures 2F-G) [10], suggesting that the amount of H_2O_2 released by gp91^{phox+} neutrophils is insufficient to induce NETs in gp91^{phox-} cells [3].

A GT approach to treat CGD may be used to overcome recalcitrant, life-threatening infections, but is currently limited as salvage therapy to experimental studies in selected patients with very poor performance status and lacking a human leukocyte antigen (HLA)-identical haematopoietic stem cell donor [4]. GT was rapidly beneficial to our CGD patient who had suffered from an otherwise incurable fungal infection. Until day +86 after GT there was no clonal dominance in bone marrow culture derived CD34$^+$ cells (not shown) nor expansion of gene-corrected cells in blood (Figure 1A). There is a risk, however, of insertional mutagenesis by transactivating retroviral vector insertions into proto-oncogenes as shown in a recent GT trial with two adult CGD patients who developed monosomy 7 and myelodysplastic syndrome (M. Grez, Institute of Biomedical Research, Georg-Speyer-Haus, Frankfurt, Germany, oral communication, April 2009) using the same gammaretroviral SF71gp91^{phox} vector [14]. Also, five patients developed leukemia in two GT trials in children with severe combined immunodeficiency [23,24]. These experiences mandate the careful follow-up of patients.

In conclusion, we show that the severe immunodeficient phenotype and the high susceptibility to *Aspergillus* infection of CGD patients might be linked to absence of NETs, and that restoration of NADPH oxidase function and NET formation by GT leads to rapid cure of refractory invasive aspergillosis in X-linked CGD.

Acknowledgments

The authors wish to thank the patient and his family for their trust. We are indebted to the medical and nursing staff of the bone marrow transplantation unit of University Children's Hospital Zurich. We would like to thank Manuel Grez and Klaus Kühlke for developing and providing the SF71gp91^{phox} vector, respectively. We are grateful to Maja Rutishauser, Corinne Wenk and Oralea Büchi for technical assistance, to Ursula Lüthi and Klaus Marquardt for electron microscopy, to Britta Laube for help with immunofluorescence imaging, to Alex Imhof for isolating *A. nidulans* conidia and to Hans Steinert for carrying out PET-CT scans.

This work was supported by a grant of the Chronic Granulomatous Disorder Research Trust, UK to J.R. and M.B., a "Forschungskredit der Universität Zürich 2006" grant to J.R. and M.B. and a grant from the "Stiftung für wissenschaftliche Forschung an der Universität Zürich/Baugarten Stiftung" to R.S. The funders had no role in study design, data collection and analysis, decision to publish, or preparation of the manuscript.

Authorship

Contribution: M.B. and A.H. performed the experiments, analyzed data and contributed to the writing of the manuscript. V.B. did the immunofluorescence image acquisition and contributed to data analysis. U.S. did the bone marrow cultures and contributed to data analysis. R.A.S. designed the clinical gene therapy protocol, attended the patient together with J.R., and contributed to writing of the manuscript. A.Z. and J.R. designed and directed the study, and contributed to the writing of the manuscript.

Conflict-of-interest disclosure: The authors declare no competing financial interests.

References

1. Brinkmann V, Reichard U, Goosmann C, et al. Neutrophil extracellular traps kill bacteria. Science. 2004;303:1532-1535.
2. Urban CF, Reichard U, Brinkmann V, Zychlinsky A. Neutrophil extracellular traps capture and kill Candida albicans yeast and hyphal forms. Cell Microbiol. 2006;8:668-676.
3. Fuchs TA, Abed U, Goosmann C, et al. Novel cell death program leads to neutrophil extracellular traps. J Cell Biol. 2007;176:231-241.
4. Seger RA. Modern management of chronic granulomatous disease. Br J Haematol. 2008;140:255-266.
5. Segal BH, DeCarlo ES, Kwon-Chung KJ, Malech HL, Gallin JI, Holland SM. Aspergillus nidulans infection in chronic granulomatous disease. Medicine (Baltimore). 1998;77:345-354.
6. Winkelstein JA, Marino MC, Johnston RB, Jr., et al. Chronic granulomatous disease. Report on a national registry of 368 patients. Medicine (Baltimore). 2000;79:155-169.
7. Bonnett CR, Cornish EJ, Harmsen AG, Burritt JB. Early neutrophil recruitment and aggregation in the murine lung inhibit germination of Aspergillus fumigatus Conidia. Infect Immun. 2006;74:6528-6539.
8. Diamond RD, Clark RA. Damage to Aspergillus fumigatus and Rhizopus oryzae hyphae by oxidative and nonoxidative microbicidal products of human neutrophils in vitro. Infect Immun. 1982;38:487-495.
9. Morgenstern DE, Gifford MA, Li LL, Doerschuk CM, Dinauer MC. Absence of respiratory burst in X-linked chronic granulomatous disease mice leads to abnormalities in both host defense and inflammatory response to Aspergillus fumigatus. J Exp Med. 1997;185:207-218.
10. Rex JH, Bennett JE, Gallin JI, Malech HL, Melnick DA. Normal and deficient neutrophils can cooperate to damage Aspergillus fumigatus hyphae. J Infect Dis. 1990;162:523-528.
11. Schaffner A, Douglas H, Braude A. Selective protection against conidia by mononuclear and against mycelia by polymorphonuclear phagocytes in resistance to Aspergillus. Observations on these two lines of defense in vivo and in vitro with human and mouse phagocytes. J Clin Invest. 1982;69:617-631.
12. Zarember KA, Sugui JA, Chang YC, Kwon-Chung KJ, Gallin JI. Human polymorphonuclear leukocytes inhibit Aspergillus fumigatus conidial growth by lactoferrin-mediated iron depletion. J Immunol. 2007;178:6367-6373.
13. Latge JP. Aspergillus fumigatus and aspergillosis. Clin Microbiol Rev. 1999;12:310-350.

14. Ott MG, Schmidt M, Schwarzwaelder K, et al. Correction of X-linked chronic granulomatous disease by gene therapy, augmented by insertional activation of MDS1-EVI1, PRDM16 or SETBP1. Nat Med. 2006;12:401-409.
15. Meshulam T, Levitz SM, Christin L, Diamond RD. A simplified new assay for assessment of fungal cell damage with the tetrazolium dye, (2,3)-bis-(2-methoxy-4-nitro-5-sulphenyl)-(2H)-tetrazolium-5-carboxanil ide (XTT). J Infect Dis. 1995;172:1153-1156.
16. Mizgerd JP. Acute lower respiratory tract infection. N Engl J Med. 2008;358:716-727.
17. Ibrahim-Granet O, Philippe B, Boleti H, et al. Phagocytosis and intracellular fate of Aspergillus fumigatus conidia in alveolar macrophages. Infect Immun. 2003;71:891-903.
18. Philippe B, Ibrahim-Granet O, Prevost MC, et al. Killing of Aspergillus fumigatus by alveolar macrophages is mediated by reactive oxidant intermediates. Infect Immun. 2003;71:3034-3042.
19. Cornish EJ, Hurtgen BJ, McInnerney K, et al. Reduced nicotinamide adenine dinucleotide phosphate oxidase-independent resistance to Aspergillus fumigatus in alveolar macrophages. J Immunol. 2008;180:6854-6867.
20. Levitz SM, Farrell TP. Human neutrophil degranulation stimulated by Aspergillus fumigatus. J Leukoc Biol. 1990;47:170-175.
21. Segal AW. How neutrophils kill microbes. Annu Rev Immunol. 2005;23:197-223.
22. Lehrer RI, Jan RG. Interaction of Aspergillus fumigatus Spores with Human Leukocytes and Serum. Infect Immun. 1970;1:345-350.
23. Hacein-Bey-Abina S, Garrigue A, Wang GP, et al. Insertional oncogenesis in 4 patients after retrovirus-mediated gene therapy of SCID-X1. J Clin Invest. 2008;118:3132-3142.
24. Howe SJ, Mansour MR, Schwarzwaelder K, et al. Insertional mutagenesis combined with acquired somatic mutations causes leukemogenesis following gene therapy of SCID-X1 patients. J Clin Invest. 2008;118:3143-3150.

Supplemental materials

Patient description and gene therapy

A 3 y old boy suffered from a severe oxygen-dependent bilateral *Aspergillus nidulans* lung infection, requiring partial resection of the left lower lung lobe. He was diagnosed with X-linked CGD (*CYBB EX1_3del*) with complete absence of gp91^{phox} protein expression and received prophylaxis with itraconazole and Co-trimoxazole. The infection was non-responsive to treatment with IV voriconazole, caspofungine, and granulocyte transfusions, requiring bone marrow transplantation. There was no human leukocyte antigen (HLA)-identical sibling or unrelated donor, and his poor clinical condition excluded transplantation from an HLA-A disparate donor.

The boy was treated with gene therapy at age 8 $^{7}/_{12}$ y with a monocistronic LTR-driven gammaretroviral SF71gp91^{phox} vector [1] under a protocol approved by the ethics review board of the University Children's Hospital Zurich and the Swiss Expert Committee for Bio-Safety after written informed consent of the parents. Collection of CD34$^+$ cells, transduction, conditioning with low-dose liposomal busulfan IV (8.8 mg/KG), and clinical follow up were performed as described [1]. Treatment with voriconazole was continued throughout and has not yet been tapered.

To increase engraftment rate [2,3], CD34$^+$ cells were transduced twice. The first batch (A) was delivered intravenously, the second one (B) by direct intra-osseous injection 6 days later. 31% (batch A) and 34.5% (batch B) of the patients CD34 cells expressed gp91 after transduction. There were 1.3 (batch A) and 0.96 (batch B) proviral copies per genome. Following conditioning, we reinfused 2.2 x10^6 (batch A) and 3.8 x10^6 (batch B) CD34gp91^{phox+} cells/KG resulting in a total dose of 6x106 transduced cells/KG. The patient experienced a short period of myelosuppression (Figure 1A). Regular follow-up is ongoing.

Methods

For all NET experiments, neutrophils were resuspended in serum free RPMI media (phenol red-free) supplemented with 10mM Hepes and used within 1 h after isolation. The same media was used to culture *A.nidulans* conidia and hyphae.

Whole body PET-CT was conducted according to standard protocols. For immunofluorescence and scanning electron microscopy (SEM) cells were processed as described [4].

References

1. Ott MG, Schmidt M, Schwarzwaelder K, et al. Correction of X-linked chronic granulomatous disease by gene therapy, augmented by insertional activation of MDS1-EVI1, PRDM16 or SETBP1. Nat Med. 2006;12:401-409.
2. Feng Q, Chow PK, Frassoni F, et al. Nonhuman primate allogeneic hematopoietic stem cell transplantation by intraosseus vs intravenous injection: Engraftment, donor cell distribution, and mechanistic basis. Exp Hematol. 2008;36:1556-1566.
3. Frassoni F, Gualandi F, Podesta M, et al. Direct intrabone transplant of unrelated cord-blood cells in acute leukaemia: a phase I/II study. Lancet Oncol. 2008;9:831-839.
4. Brinkmann V, Reichard U, Goosmann C, et al. Neutrophil extracellular traps kill bacteria. Science. 2004;303:1532-1535.

Supplemental figure 1

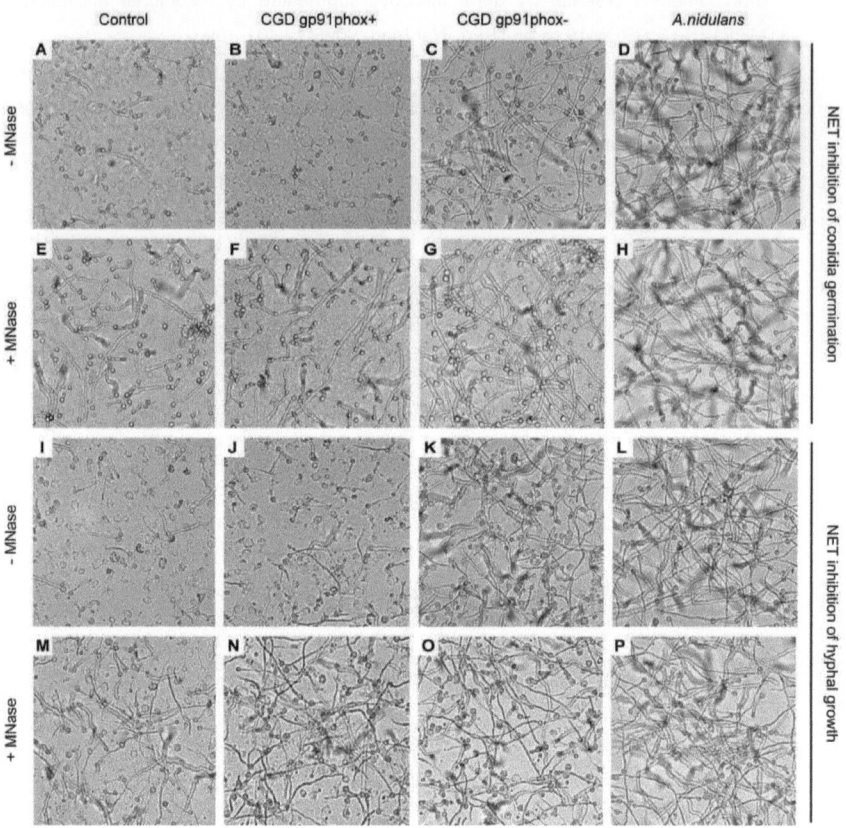

Figure S1. NET antifungal activity on *A. nidulans* conidia and hyphae. NET control of *A. nidulans* conidia (panels A-H) and hyphae (panels I-P). Conidia were plated on FACS-sorted neutrophils stimulated with PMA for 4 h, i.e. until NET formation was complete, cells were dead and therefore incapable of phagocytosis. Hyphae were co-incubated with FACS-sorted neutrophils and PMA. Germination and growth of conidia were analyzed after 16 h incubation at 37°C; growth of hyphae after 5 h, revealing NET antifungal activity of control (A, I) and CGD gp91^{phox+} (B, J) neutrophils. This activity was significantly reduced after digestion of NETs with Micrococcal Nuclease (MNase) (E, M and F, N). CGD gp91^{phox-} neutrophils were inefficient in controlling fungal growth (C, K). *A. nidulans* growth without addition of cells (D-P). *A. nidulans* survival was visualized as hyphal outgrowth and pictures were taken in parallel to the other experiments shown in Figure 2 using an inverted light microscope coupled to a CCD camera. All experiments were done at MOI 0.5 (*A. nidulans* : neutrophils) and were repeated at least three times with similar results.

Supplemental figure 2

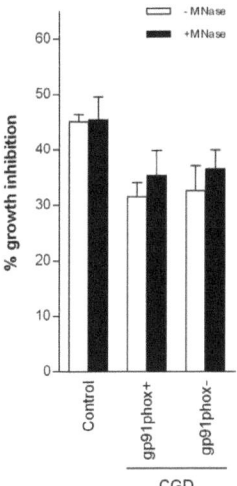

Figure S2. NET independent inhibition of *A. nidulans* hyphal growth. Hyphal outgrowth was measured after 16 h pre-incubation of conidia in media. Hyphae were co-incubated with FACS-sorted neutrophils and PMA ± MNase and hyphal viability was assessed after 2h (i.e. before NETs formation) at 37°C, revealing NADPH oxidase independent growth inhibition during this time. The data shown are presented as mean ± SD of a representative triplicate experiment. Inhibition of fungal growth is expressed as percentage of control values (*A. nidulans* hyphae incubated in media).

Cover illustration of *Blood*, 24 September 2009, volume 114

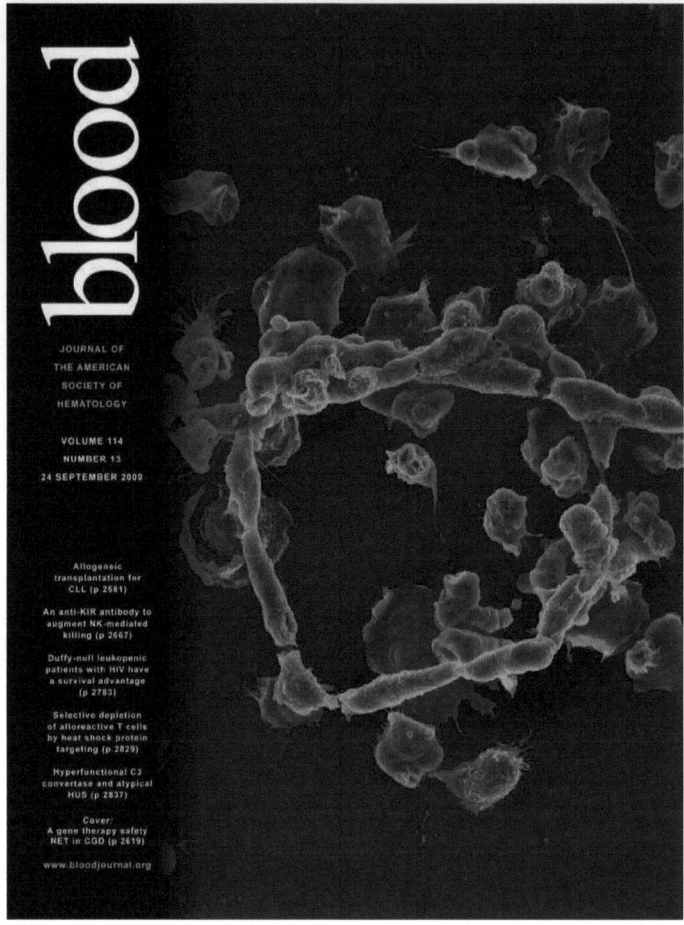

Cover caption: It is unclear how neutrophils control *Aspergillus* species in healthy persons. Due to their large size, *Aspergillus* hyphae (red) cannot be phagocytosed by neutrophils (green). Bianchi et al show in this issue that recently discovered NADPH oxidase-dependent microbicidal pathways through neutrophil extracellular traps (NETs) are efficient against *Aspergillus* conidia and hyphae *in vitro*, and that restoration of NET formation by gene therapy of NADPH oxidase-deficient X-linked chronic granulomatous disease aided clearing severe invasive *Aspergillus nidulans* infection *in vivo*, underlining the role of functional NADPH oxidase in NET formation and antifungal activity. See the article by Bianchi et al on page 2619.

Letter sent to *Blood* regarding the paper: Bianchi M, et al. Restoration of NET formation by gene therapy in CGD controls aspergillosis. *Blood* 2009;114:2619–2622

CORRESPONDENCE

Reconstitution of protection against *Aspergillus* infection in chronic granulomatous disease (CGD)

Quinten Remijsen[1], Peter Vandenabeele[2], Jean Willems[1], and Taco W. Kuijpers[3]

[1] Laboratory of Biochemistry, Department of Medicine, KU Leuven Campus Kortrijk, Kortrijk, Belgium

[2] Molecular Signaling and Cell Death Unit, Department for Molecular Biomedical Research VIB, Ghent University, Ghent, Belgium

[3] Department of Blood Cell Research, Sanquin Research and Landsteiner Laboratory, Academic Medical Center, University of Amsterdam, Emma Children's Hospital, Amsterdam, The Netherlands

Corresponding Author:

Quinten Remijsen

KU Leuven campus Kortrijk, IRC

Laboratory of Biochemistry

E Sabbelaan 53, 8500 Kortrijk, Belgium

e-mail: quinten.remijsen@kuleuven-kortrijk.be

CORRESPONDENCE

To the editor:

We would like to comment on the recent paper by Bianchi andcoworkers.[1] Patients diagnosed with chronic granulomatous disease (CGD) suffer from recurrent and life-threatening infections. Analysis of clinical data from 429 European CGD patients recently confirmed that *Aspergillus* is a major threat, causing mostly pneumonia but also brain abscesses.[2]

Reconstitution of an intact X-linked gene encoding gp91phox is a remarkable accomplishment and a promising new therapy for X-CGD. In their Blood paper, Bianchi et al correlate the reconstituted nicotinamide adenine dinucleotide phosphate (NADPH) oxidase activity in this patient with protection against *Aspergillus* infection, as a result of reconstituted neutrophil extracellular trap (NET) formation. NETs are composed of chromatin decorated with granular proteins and bind Gram-positive and -negative bacteria, as well as fungi, in a nonspecific way. Recently, *ex vivo* NET formation was shown to depend on superoxide production by activated NADPH oxidase.[3] The only empirical data provided by Bianchi and coauthors to underscore the crucial role of NET formation against *Aspergillus* infection are *ex vivo* observations showing a microbicidal effect of NETs against *Aspergillus* species.

Treating the *ex vivo* findings as proof of a causal link between (reconstituted) NET formation and protection against *Aspergillus* may be a step too far and may not be essential. NETs do not have any specificity in microbicidal activity, whereas CGD patients show a selective susceptibility only to certain microbes.

Moreover, *Aspergillus* infection has recently been studied in a CGD mouse model, and activity of indoleamine 2,3-dioxygenase (IDO) was shown to be crucial for the survival of *Aspergillus* infection.4 IDO converts L-tryptophan into L-kynurenine but requires superoxide as a cofactor for its activity. Secreted L-kynurenine subsequently acts as an anti-inflammatory agent by mechanisms that are incompletely understood but have been shown to induce cell death in proinflammatory $\gamma\delta$ T-cell subsets producing interleukin-17.[4] Romani et al thus concluded that hyperinflammation caused a lethal outcome for CGD mice upon challenge with *Aspergillus*, rather than a defective clearance itself, as was previously suggested in CGD patients with an overwhelming pulmonary aspergillosis.[5] This seems plausible in view of a large body of evidence showing that CGD patients often display exaggerated immune responses against immunologic challenges,[6] and granulomas have been shown to develop in the absence of any detectable pathogen, even after wound sterilization or after injection of heat-inactivated pathogens.[7]

Romani and coworkers underscore their conclusion by demonstrating that CGD mice, which all died upon *Aspergillus* challenge, survive this infection when treated with the IDO product L-kynurenine in combination with interferon-γ. In turn, wild-type mice, which normally survive this challenge, no longer overcome *Aspergillus* infection when treated with the IDO inhibitor 1-methyl tryptophan.[8] Although the specificity of the challenge and the background of the CGD mouse strain are also subject for debate, the abovementioned relevant findings were not discussed or referred to in the Bianchi paper. It might be possible that reconstitution of other superoxide-dependent steps, such as direct effects or indirectly via IDO activity, rather than the restoration of the capability of NET formation, will protect CGD patients from *Aspergillus* infection.

Conflict-of-interest disclosure: The authors declare no competing financial interests.

References

1. Bianchi M, Hakkim RA, Brinkmann V, et al. Restoration of NET formation by gene therapy in CGD controls aspergillosis. Blood. 2009;114:2619–2622.
2. van den Berg JM, van Koppen E, Ahlin A, et al. Chronic granulomatous disease: the European experience. PLoS ONE. 2009;4:e5234.
3. Fuchs TA, Abed U, Goosmann C, et al. Novel cell death program leads to neutrophil extracellular traps. J Cell Biol. 2007;176:231–241.
4. Romani L, Fallarino F, De Luca A, et al. Defective tryptophan catabolism underlies inflammation in mouse chronic granulomatous disease. Nature. 2008;451:211–215.
5. Siddiqui S, Anderson VL, Hilligoss DM, et al. Fulminant mulch pneumonitis: an emergency presentation of chronic granulomatous disease. Clin Infect Dis. 2007;45:673–681.
6. Bylund J, Macdonald KL, Brown KL, et al. Enhanced inflammatory responses of chronic granulomatous disease leukocytes involve ROS-independent activation of NF-kappa B. Eur J Immunol. 2007;37:1087–1096.
7. Morgenstern DE, Gifford MA, Li LL, Doerschuk CM, Dinauer MC. Absence of respiratory burst in X-linked chronic granulomatous disease mice leads to abnormalities in both host defense and inflammatory response to Aspergillus fumigatus. J Exp Med. 1997;185:207–218.
8. Grohmann U, Volpi C, Fallarino F, et al. Reverse signaling through GITR ligand enables dexamethasone to activate IDO in allergy. Nat Med. 2007;13:579–586.

Response sent to *Blood* regarding the correspondence: Remijsen Q, et al. Reconstitution of protection against *Aspergillus* infection in chronic granulomatous disease (CGD). *Blood* 2009;114:3497

CORRESPONDENCE

Response: Protecting against *Aspergillus* infection in CGD

Abdul Hakkim[1], Robert Hurwitz[2], Matteo Bianchi[3], Volker Brinkmann[4], Ulrich Siler[3], Reinhard A. Seger[3], Arturo Zychlinsky[1], and Janine Reichenbach[3]

[1] Department of Cellular Microbiology, Max Planck Institute for Infection Biology, Berlin, Germany
[2] Protein Facility, Max Planck Institute for Infection Biology, Berlin, Germany
[3] Division of Immunology/Hematology/BMT, University Children's Hospital, Zurich, Switzerland
[4] Microscopy Core Facilities, Max Planck Institute for Infection Biology, Berlin, Germany
[5] Department of Cellular Microbiology, Max Planck Institute for Infection Biology, Berlin, Germany

Corresponding Author:
Janine Reichenbach
University Children's Hospital Zurich
Division of Immunology/Hematology/BMT
Steinwiesstrasse 75, Zurich 8032, Switzerland
e-mail: janine.reichenbach@kispi.uzh.ch

Published in: *Blood* 2009;114:3498
Own contribution: M.B. contributed to the experiments and to the writing of the correspondence.

CORRESPONDENCE

Response:

We would like to thank Remijsen and colleagues for their comments regarding our recent publication in *Blood* titled "Restoration of NET formation by gene therapy in CGD controls aspergillosis."[1]

We regret that we neglected to cite the work of Romani et al.[2] This was because the "Brief Report" format in *Blood* is restricted in length and number of citations. Romani et al reported that superoxide produced by the nicotinamide adenine dinucleotide phosphate (NADPH) oxidase regulates indoleamine 2,3-dioxygenase (IDO), which leads to L-kynurenine formation and eventually to the activation of interleukin-17 (IL-17)–producing T cells, which are involved in the acquired immune response. Indeed, treatment with L-kynurenine and interferon-γ helps resolve *Aspergillus* infections in p47phox-deficient mice, a murine model of chronic granulomatous disease (CGD).

We analyzed the level of L-kynurenine and IL-17 in sera collected from the CGD patient described in our publication[1] before, as well as 11, 17, and 98 days after, gene therapy (GT). There were no significant differences in L-kynurenine levels in the sera before and after GT as measured by high-performance liquid chromatography.[3] The concentration of IL-17 in sera was less than the detection limit of 30 pg/ml of the enzyme-linked immunosorbent assay (R&D Systems) at all time points analyzed. Because of ethical considerations, we did not analyze local L-kynurenine and IL-17 levels in tissue biopsies, and therefore cannot exclude a local effect of the IDO pathway on *Aspergillus* infection.

The complex clinical phenotype in CGD patients reflects the pleiotropic functions of the NADPH oxidase in immune defense. It is likely that several mechanisms, including the activity of IDO, contribute to the restoration of immune defense after GT, and it would certainly be important to address this point if the opportunity arises. Treatment with L-kynurenine, however, has not yet been tested in CGD patients because of concerns about its epileptogenic potential.

The high incidence of aspergillosis compared with other opportunistic infections in CGD patients is indeed difficult to explain. It would certainly be interesting to survey how neutrophils isolated from these patients respond to a spectrum of different opportunistic microbes. Interestingly, knockout mice in any of the NADPH oxidase subunits seem susceptible to diverse microbial insults. Regardless, it is clear that the NADPH oxidase, as a part of the innate immune system, is initially important in microbe clearance, through, for example, phagocytosis and neutrophil extracellular trap (NET) formation. This enzyme, as elegantly shown by Romani et al, is also important in the second function of innate immunity, which is to set the stage for an acquired immune response.

We reported the acquisition of in vitro anti-*Aspergillus* activity of NETs after GT. This is in line with the observations that fungi are more susceptible to NETs than to phagocytic killing.[4] More importantly, the patient started to get better within few days after engraftment of gene-transduced cells, suggesting an immediate innate response. The role of NETs in the recovery of the patient will remain a correlation since we reported the *in vitro* function of human cells; *in vivo* experiments are obviously out of the question. It is likely that the improvement after restoration of NADPH oxidase activity in the CGD patient reported was due to several pathways where this enzyme is involved. Our data, however, suggest that NETs played a prominent role in the clearance of the *Aspergillus* infection.

Approval was obtained from the ethics review board of the University Children's Hospital Zurich and the Swiss Expert Committee for Bio-Safety for these studies. Informed consent was provided according to the Declaration of Helsinki.

Conflict-of-interest disclosure: The authors declare no competing financial interests.

References

1. Bianchi M, Hakkim A, Brinkmann V, et al. Restoration of NET formation by gene therapy in CGD controls aspergillosis. Blood. 2009;114:2619–2622.
2. Romani L, Fallarino F, De Luca A, et al. Defective tryptophan catabolism underlies inflammation in mouse chronic granulomatous disease. Nature. 2008;451:211–215.
3. Widner B, Werner ER, Schennach H, Wachter H, Fuchs D. Simultaneous measurement of serum tryptophan and kynurenine by HPLC. Clin Chem. 1997;43:2424–2426.
4. Urban CF, Reichard U, Brinkmann V, Zychlinsky A. Neutrophil extracellular traps capture and kill Candida albicans yeast and hyphal forms. Cell Microbiol. 2006;8:668–676.

CHAPTER 2

Restoration of anti-*Aspergillus* defense by neutrophil extracellular traps in human chronic granulomatous disease after gene therapy is calprotectin-dependent

Matteo Bianchi, MSc,[a*] Maria J. Niemiec, MSc,[b*] Ulrich Siler, PhD,[a] Constantin F. Urban, PhD,[b‡] and Janine Reichenbach, MD,[a‡]

[*,‡] These authors contributed equally to this work

[a] Division of Immunology/Hematology/BMT, University Children's Hospital Zurich, Switzerland

[b] Antifungal Immunity Group, Molecular Biology Department, Laboratory for Molecular Infection Medicine Sweden, Umeå University, Umeå, Sweden

Corresponding Authors:

Janine Reichenbach
Division of Immunology/Hematology/BMT
University Children's Hospital Zurich
Steinwiesstrasse 75, 8032 Zurich, Switzerland
Phone: +41 44 266 7311, Fax: +41 44 266 7914
e-mail: janine.reichenbach@kispi.uzh.ch

Constantin F. Urban
Antifungal Immunity Group, Molecular Biology Department
Laboratory for Molecular Infection Medicine Sweden
Umeå University
90187 Umeå, Sweden
Phone: +46 907850806, Fax: +46 90772630
e-mail: constantin.urban@molbiol.umu.se

Published in: *Journal of Allergy and Clinical Immunology* 2011;127:1243-52

Own contribution: M.B. performed the experiments (except Fig.3E), analyzed data, prepared all figures and contributed to the writing of the manuscript.

Abstract

Background: *Aspergillus spp* infection is a potentially lethal disease in patients with neutropenia or impaired neutrophil function. We showed previously that *Aspergillus* hyphae which are too large for neutrophil phagocytosis are inhibited by reactive oxygen species-dependent neutrophil extracellular trap (NET) formation. This process is defective in chronic granulomatous disease (CGD) due to impaired phagocyte nicotinamide adenine dinucleotide phosphate (NADPH) oxidase function.

Objective: To determine the antifungal agent and mechanism responsible for reconstitution of *Aspergillus* growth inhibition within NETs after complementation of NADPH oxidase function by gene therapy (GT) for CGD.

Methods: Antifungal activity of free and NET-released calprotectin was assessed by incubation of *A. nidulans* with purified calprotectin, induced NETs from human controls and CGD neutrophils after GT in presence or absence of Zn^{2+} or α-S100A9 antibody, and with induced NETs from wild type or $S100A9^{-/-}$ mouse neutrophils.

Results: We identified the host Zn^{2+} chelator calprotectin as neutrophil-associated antifungal agent expressed within NETs, reversibly preventing *A. nidulans* growth at low concentration, and leading to irreversible fungal starvation at higher concentration. Specific antibody-blocking and Zn^{2+} addition abolished calprotectin-mediated inhibition of *A. nidulans* proliferation *in vitro*. The role of calprotectin in anti-*Aspergillus* defence was confirmed in calprotectin knockout mice.

Conclusion: Reconstituted NET-formation by GT for human CGD was associated with rapid cure of pre-existing therapy refractory invasive pulmonary aspergillosis *in vivo*, underlining the role of functional NADPH oxidase in NET-formation and calprotectin release for antifungal activity. These results demonstrate the critical role of calprotectin in human innate immune defense to *Aspergillus* infection.

Capsule Summary

Susceptibility to *Aspergillus* infection of patients with CGD might be linked to absence of NADPH oxidase-dependent NETs. Restoration of NET-formation and calprotectin-release by gene therapy leads to rapid cure of refractory invasive aspergillosis in CGD.

Key Words

Chronic granulomatous disease, gene Therapy, neutrophil extracellular trap, calprotectin, *Aspergillus* infection.

Abbreviations

CGD: Chronic granulomatous disease
DAPI: 4',6-diamidino-2-phenylindole
GT: Gene therapy
LTR: Long terminal repeat
NADPH: Nicotinamide adenine dinucleotide phosphate
NE: Neutrophil elastase
NET: Neutrophil extracellular trap
PMA: Phorbol-12-myristate-13-acetate
ROS: Reactive oxygen species

Introduction

Neutrophils kill microbes by distinct reactive oxygen species (ROS) dependent processes: intracellularly following phagocytosis and by extracellular mechanisms including neutrophil extracellular traps (NETs), which capture and kill bacteria [1], parasites [2] and fungi [3-4]. NETs are composed of chromatin (histones and DNA), decorated with at least 20 other proteins [1, 4-5]. In naïve neutrophils 8 of these proteins localize to granules, such as neutrophil elastase, myeloperoxidase or azurocidin, and 11 to the cytoplasm, such as catalase and the calprotectin complex [4]. This heteroduplex formed by subunits S100A8 and S100A9 belongs to the calcium-binding protein family of S100 proteins, and exerts strong microbistatic activity against a variety of microorganims *in vitro* [6].

Defective nicotinamide adenine dinucleotide phosphate (NADPH) oxidase function accounts for impaired phagocyte ROS production and poor microbial killing in chronic granulomatous disease (CGD), leading to recurrent life-threatening bacterial and fungal infections. *Aspergillus spp* infections cause pneumonia and disseminated disease and are the leading cause of death in these patients [7-10]. ROS production by the NADPH oxidase is indispensable for microbe-triggered NET formation [5, 11-12], whilst their mechanism of action in this process is unknown. After release from granules neutrophil elastase (NE) and myeloperoxidase (MPO) synergize to drive decondensation of chromatin, an early event preceding the release of NETs [13-14]. ROS might contribute to the release of NE and MPO from granules. We previously reported impaired NET-formation in CGD *in vitro*, resulting in defective clearance of *Aspergillus* infection *in vivo*. NADPH oxidase-dependent ROS production and NET-inhibition of *Aspergillus* conidia and hyphae were restored early after gene therapy (GT) in a patient with X-linked gp91^{phox} (phox, *p*hagocyte *ox*idase)-deficient CGD, paralleled by clearance of therapy refractory *Aspergillus nidulans* lung infection [11].

The antifungal agent responsible for *A. nidulans* growth inhibition within NETs has not been characterized. Here we show that amongst the 24 NET associated proteins the major antifungal agent inhibiting *Aspergillus* growth within NETs is the cytoplasmic calprotectin protein complex (S100A8/A9 heteroduplex). This is in good agreement with our previous report showing that calprotectin is the major NET component against *Candida albicans* [4].

Methods

We treated an 8 7/12 years old boy with X-linked gp91^{phox}-deficient CGD and multifocal therapy refractory *A. nidulans* lung infection with a monocistronic long terminal repeat (LTR)-driven gammaretroviral SF71gp91^{phox} vector (see reference [11] for details; additional information in the "Methods" section in the Online Repository) under a protocol approved by the ethics review board of the University Children's Hospital Zurich and the Swiss Expert Committee for Bio-Safety after written informed consent of the parents.

NET induction and quantification

NET formation was quantified as described [11] after stimulation of 5x10^4 neutrophils (control, CGD unsorted, CGD gp91^{phox}-negative (gp91^{phox-}) and -positive (gp91^{phox+}) sorted neutrophils) for 4h with 40nM phorbol-12-myristate-13-acetate (PMA) and staining the NET-DNA with 1 µM Sytox green (Invitrogen) in a black 96-well plate. Non-stimulated neutrophils were used as negative control. The plates were read in a fluorescence microplate reader (Victor3, PerkinElmer) with a filter setting of 485/535 nm (excitation/emission).

A. nidulans growth inhibition

Antifungal activity of purified calprotectin was assessed incubating *A. nidulans* conidia or hyphae with 0.5-128 µg/ml recombinant S100A8 and S100A9 (ProtEra) ± 1 µM ZnSO$_4$ or 15 µg/ml rabbit polyclonal α-S100A9 antibody (H00006280-D01P, Abnova). Antifungal activity of NETs was determined incubating *A. nidulans* conidia or hyphae (strain isolated as described [15] from bronchoalveolar lavage fluid of the patient) with human and murine NETs or human NET-extracts ± 0 to 2 µM ZnSO$_4$ or 15 µg/ml rabbit polyclonal α-S100A9 antibody, respectively (additional Methods information in the Online Repository).

NET induction with *A. nidulans*

In a 24-well plate, 2x10^5 control, CGD gp91^{phox-} and gp91^{phox+} neutrophils were incubated with *A. nidulans* hyphae obtained from culture of 6x10^4 *A. nidulans* conidia on polylysine coated cover slips for 5 h at 37 °C, then fixed in 4% paraformaldehyde for bright field microscopy and immunostaining/confocal microscopy, or 2.5% glutaraldehyde for scanning electron microscopy (SEM). For confocal microscopy (SP5, Leica) specimens were blocked with 5% donkey serum (Jackson), 1% BSA, 3% cold water fish gelatine (Sigma-Aldrich) and 0.25% Tween 20 in PBS, incubated with primary antibodies directed against S100A8/A9 (BM4029 and BM4027, Acris), common *Aspergillus spp* soluble proteins (NB100-65026, Novus Biologicals), and species-specific secondary antibodies coupled to Alexa Fluor 488 and 568 (Invitrogen). DNA was stained with 4',6-diamidino-2-phenylindole (DAPI, Sigma-

Aldrich). For SEM (CM208, Philips), specimens were postfixed with 2% osmium tetroxid, dehydrated with graded ethanol series (50 to 100%), critical-point dried, and coated with 10 nm platinum.

Release of calprotectin

In a 96-well plate, 5×10^4 control, CGD gp91^{phox-} and gp91^{phox+} neutrophils were stimulated in duplicate with 40 nM PMA at 37°C and 5% CO_2 for 4 h to form NETs. After 30 min NET digestion with 5 U/ml micrococcal nuclease (MNase) + 1 U/ml deoxyribonuclease (DNase)-1 samples were centrifuged 10 min at 10'000 g to remove debris. Heterodimeric calprotectin (S100A8/A9) concentration was measured by ELISA (Hycult, detection limit 1.6 ng/ml) with 1/10, 1/40 and 1/80 dilutions. As control, total cytoplasmic calprotectin was quantified after neutrophil lysis.

Results

NET quantification in CGD neutrophils after GT

We aimed to identify factors that limit A. nidulans growth in NADPH oxidase-dependent NETs. Therefore we first analyzed whether NET-formation was restored 2.6 years after GT in gp91^{phox+} (transduced) compared to gp91^{phox-} (non-transduced) neutrophils from the reported CGD patient [11]. Staining with α-gp91^{phox} antibody showed 22.8% gp91^{phox+} neutrophils. Sorting of gp91^{phox-} and gp91^{phox+} neutrophils by autoMACS (Miltenyi Biotec) resulted in purity of 98.3% and 92.5% (Fig 1, A).

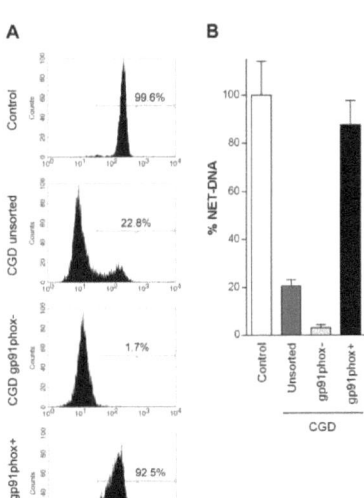

FIG 1. NET quantification in CGD gp91^{phox+} and gp91^{phox-} sorted neutrophils. A, Neutrophils were sorted by magnetic beads and autoMACS. B, NET-formation was quantified after PMA activation of neutrophils, by staining of released NET-DNA with Sytox green. Data are mean ± SD of representative triplicate experiments. Difference between control and gp91^{phox+} neutrophils was non-significant ($P > 0.05$).

Quantification of NET-formation in CGD neutrophils after GT correlated well with the percentage of gp91^{phox+} cells, 20.5% of unsorted, 3.2% of gp91^{phox-}, and 87.6% of gp91^{phox+} neutrophils releasing NETs (Fig 1, B). The O$_2^-$ production of individual gp91^{phox+} neutrophils was 27.8% (compared to control), measured by reduction of cytochrome c (data not shown) [16], indicating this threshold as enough for substantial, albeit not 100% NET-formation.

Calprotectin inhibits *A. nidulans* growth efficiently

We previously identified the cytoplasmic protein complex calprotectin as major antifungal effector in NETs preventing growth of *C. albicans* [4] and therefore reasoned that calprotectin might also be responsible for NET-mediated growth inhibition of *A. nidulans*. There is currently no information available whether purified calprotectin is able to inhibit *Aspergillus spp.* To evaluate the inhibitory activity of each calprotectin subunit on *A. nidulans* growth, we incubated conidia and hyphae with purified subunits S100A8 and S100A9 [17]. Addition of S100A8 did not inhibit fungal growth, whereas S100A9 had a small effect on conidia germination, but not on hyphae growth (Fig 2, A-B and E1 in the Online Repository). Only both subunits together strongly inhibited *A. nidulans* growth, indicating that the heteroduplex is required for full inhibition of *Aspergillus spp*, consistent with a report on its anti-candidal activity [18]. In all cases, addition of 1 µM zinc (Zn^{2+}) restored fungal growth, confirming the putative role of Zn^{2+} chelation by calprotectin in suppressing fungal growth [18-20].

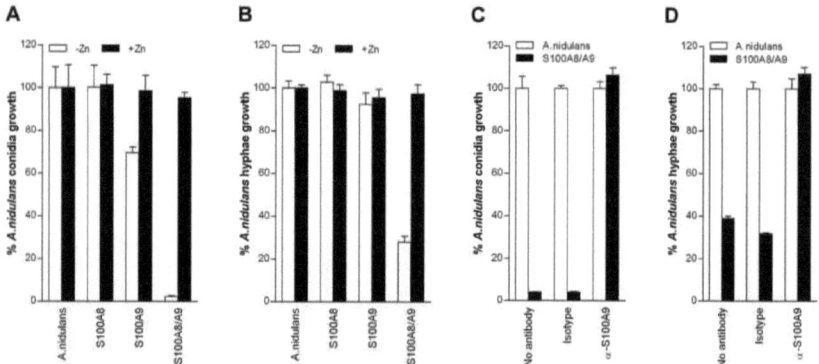

FIG 2. Inhibition of *A. nidulans* growth by purified calprotectin. A-B, Conidia and hyphae incubated with 5 µg/ml S100A8 and/or S100A9 ± 1 µM Zn^{2+}. C-D, Conidia and hyphae incubated with 5 µg/ml S100A8 and S100A9 ± 15 µg/ml polyclonal α-S100A9 antibody or unspecific control antibody. Data are mean ± SD of representative triplicate experiments.

Additionally, we pre-incubated combined S100A9 and S100A8 with a polyclonal α-S100A9 antibody. *A. nidulans* growth of conidia and hyphae was restored in the presence of specific antibody (Fig 2, C-D and E2), confirming the requirement for interaction of both calprotectin subunits for antifungal activity. Unspecific control antibody did not have any effect. Here we used an excess amount of 5 µg/ml purified S100A8 and S100A9, but similar inhibitory effects were obtained with as low as 0.5 µg/ml, a calprotectin amount found in NETs from about 4 million neutrophils [4].

Antifungal activity of CGD gp91^{phox+} cells is calprotectin dependent

To study whether restored antifungal activity of CGD gp91^{phox+} neutrophils after GT was attributable to calprotectin, we PMA-stimulated gp91^{phox+} and gp91^{phox-} neutrophils to induce NETs, and coincubated with A. nidulans conidia and hyphae ± Zn^{2+}. Only gp91^{phox+} neutrophils showed strong antifungal activity, equivalent to control (Fig 3, A-B and E3). gp91^{phox-} neutrophils did not form NETs (Fig 1, B), yet partially inhibited conidia germination, probably by phagocytosis [21], but were inefficient against hyphae. Full restoration of A. nidulans growth by addition of Zn^{2+} supports the assumption that NET antifungal activity was calprotectin dependent (at calprotectin concentrations of 450-500 ng/ml, measured by ELISA).

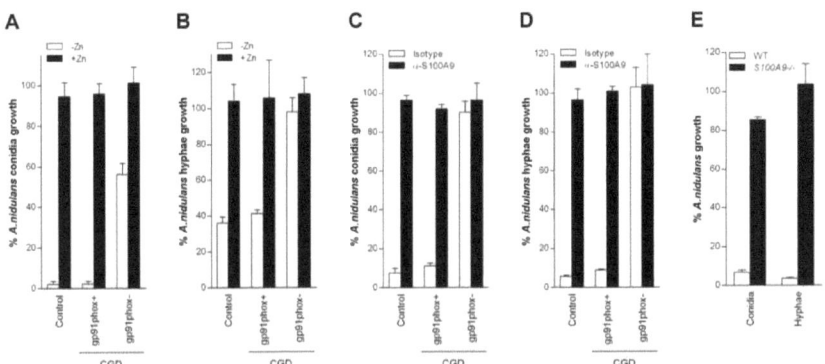

FIG 3. Calprotectin dependent antifungal activity of CGD gp91^{phox+} neutrophils. A. nidulans conidia or hyphae incubated with NETs (A-B) or NET-supernatant (C-D) from control and CGD sorted neutrophils ± 1 µM Zn^{2+} or 15 µg/ml α-S100A9, respectively. E, Wild type and S100A9$^{-/-}$ mouse NETs infected with A. nidulans conidia or hyphae. Data are mean ± SD of representative triplicate experiments.

Restoration of A. nidulans conidia and hyphae growth by Zn^{2+} addition was dose dependent (Fig E4). Addition of Mn^{2+}, Fe^{2+} and Cu^{2+} (0 to 2.5 µM) however, did not restore fungal growth (not shown); indicating that Zn^{2+} binding by calprotectin has a major NET anti-Aspergillus activity. Since addition of metal ions can have pleiotropic effects, we applied an α-S100A9 antibody to block calprotectin-mediated inhibition in concentrated NET-extracts, subsequently incubated with A. nidulans conidia or hyphae. NET-extracts from gp91^{phox+} neutrophils in presence of unspecific control antibody showed strong antifungal activity, similar to control (Fig 3, C-D and E5). Supernatants from non-NET forming gp91^{phox-} neutrophils in presence of unspecific control antibody did not have antifungal activity, even against conidia germination, supporting the hypothesis that their low antifungal activity observed against conidia (Fig 3, A) was due to phagocytosis.

We then tested if NET-mediated A. nidulans inhibition was calprotectin dependent comparing wild-type and calprotectin-deficient mouse neutrophils. NETs were PMA-induced in mature neutrophils from both mouse strains and incubated with A. nidulans conidia and hyphae. Indeed, NETs from calprotectin-deficient neutrophils did not prevent A. nidulans growth, whereas NETs from wild-type neutrophils did (Fig 3, E), although the amount of NET-formation was similar in calprotectin-deficient and wild-type neutrophils [4]. Complete restoration of fungal growth in presence of α-S100A9 antibody in human gp91^{phox+} and control NET-extracts and the inability of calprotectin-deficient murine NETs to block fungal growth confirmed the essential antifungal role of NET-associated calprotectin against A. nidulans.

Antifungal effect of calprotectin

To determine whether calprotectin-mediated inhibition of A. nidulans growth within NETs was reversible or irreversible we incubated conidia and hyphae on NETs for 4 days to 4 weeks at neutral pH, then added Zn^{2+} to restore fungal growth. NETs strongly inhibited A. nidulans conidia germination and hyphae growth even after 4 weeks (Fig 4, A-P), indicating strong stability of calprotectin/Zn^{2+} complexes, agreeing well with their reported stability against proteases [22]. Surprisingly, addition of Zn^{2+} restored fungal growth even after 4 weeks, demonstrating that calprotectin has fungistatic, but not fungicidal activity on A. nidulans at low concentrations (450-500 ng/ml, measured by ELISA).

We next tested whether antifungal activity was still reversible at higher calprotectin concentrations which might occur *in vivo*. We incubated conidia and hyphae for 4 days with increasing concentrations of purified calprotectin S100A8 and S100A9, then added Zn^{2+} to restore fungal growth. Conidia germination was inhibited at any concentration (Fig 4, Q, white bars) whereas hyphae were slightly more resistant showing a dose-dependent increase of growth arrest between 0.5 and 4 µg/ml of S100A8/A9 (Fig 4, R, white bars). Fungal growth was not restored by Zn^{2+} when S100A8/A9 concentration was higher than 16 µg/ml (Fig 4, Q-R, black bars). Light microscopy showed intact hyphae at low S100A8/A9 concentrations (Fig 4, S), whereas in presence of >16 µg/ml S100A8/A9 hyphae had a translucent cell wall and a fragmented internal structure (Fig 4, T). Thus, we demonstrate a previously unknown microbicidal effect against A. nidulans at high calprotectin concentrations.

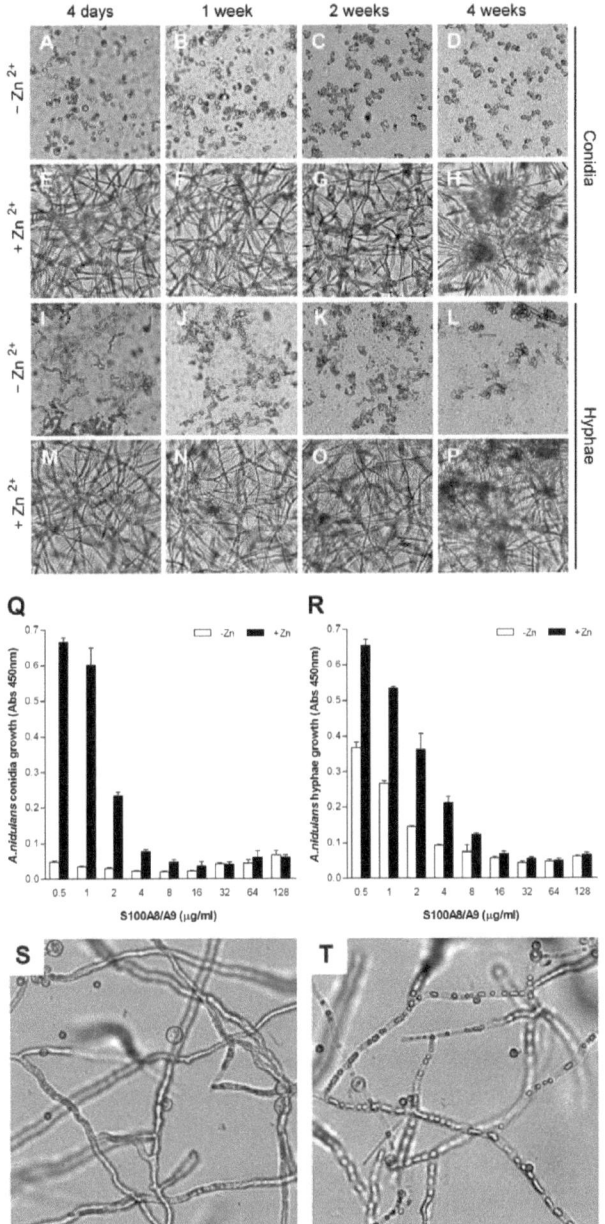

FIG 4. Antifungal effect of calprotectin. A-P, Fungistatic activity: *A. nidulans* conidia or hyphae incubated with NETs for 4 days to 4 weeks, after which 1 µM Zn^{2+} was added overnight. Q-R, Fungicidal activity: *A. nidulans* conidia or hyphae incubated with 0.5 (S) to 128 (T) µg/ml purified S100A8/A9 for 4 days, after which 50 µM Zn^{2+} was added overnight.

A. nidulans-induced NET-formation is strictly dependent on NADPH oxidase

We investigated whether the induction of NETs by *A. nidulans* is dependent on functional NADPH oxidase and is restored after CGD GT. CGD gp91^{phox+} and gp91^{phox-} neutrophils were incubated with *A. nidulans* hyphae and observed for NET-formation by light microscopy, SEM, and by immunostaining with confocal microscopy. Hyphae induced NETs in control and gp91^{phox+} neutrophils but, as expected, less strongly compared to PMA stimulation. We observed that NET formation mainly occurred when neutrophils were in direct contact with hyphae (Fig 5, *A-B*, black arrows). This might suggest that close pathogen-neutrophil interaction promotes induction of NETs. The molecular details behind this trigger mechanism will be subject to further investigation. In contrast, gp91^{phox-} neutrophils did not form NETs when in contact with hyphae (Fig 5, *C*, white arrowheads), underlining the role of functional NADPH oxidase in NET-formation and antifungal defense. Analysis by SEM showed that *A. nidulans* hyphae were entangled by NETs from control and gp91^{phox+} neutrophils (Fig 5, *D, G* and *E, H*), whereas gp91^{phox-} neutrophils surrounded and wrapped around hyphae but did not make NETs (Fig 5, *F, I*).

Calprotectin is NET-associated after *A. nidulans* incubation

The association of calprotectin and *A. nidulans*-induced NETs was studied by immunofluorescence and confocal microscopy, staining DNA with DAPI and calprotectin and hyphae with primary antibodies. Colocalization of DNA (blue) and calprotectin (red) demonstrates that calprotectin was released and associated with NETs when control (Fig 5, *J, M, P*) and gp91^{phox+} (Fig 5, *K, N, Q*) neutrophils were incubated with *A. nidulans* hyphae (green). Moreover, we demonstrate that gp91^{phox-} neutrophils neither showed NET-formation nor extracellular calprotectin upon *A. nidulans* incubation (Fig 5, *L, O, R*), although they were able to recognize *A. nidulans*, as judged by migration towards and wrapping around hyphae (Fig 5, *R*). DNA was confined to the nucleus (Fig 5, *L*) and calprotectin to the cytoplasmic compartment of these cells (Fig 5, *O*).

Notably, the quantification of calprotectin by ELISA confirmed our microscopical findings. Control, gp91^{phox+} and gp91^{phox-} neutrophils were activated with PMA for 4h and calprotectin release was determined by ELISA after NET-formation and subsequent digestion of NETs. Control and gp91^{phox+} neutrophils released approximately four times the calprotectin concentration compared to gp91^{phox-} neutrophils (Fig 6), supporting the previous results. Differences in total cytoplasmic calprotectin concentration between control, gp91^{phox+} and gp91^{phox-} neutrophils were not statistically significant (data not shown). Thus, we conclude that NADPH oxidase is not only required for the release of NETs but also for the release of calprotectin, the crucial component for the inhibition of *A. nidulans* hyphae.

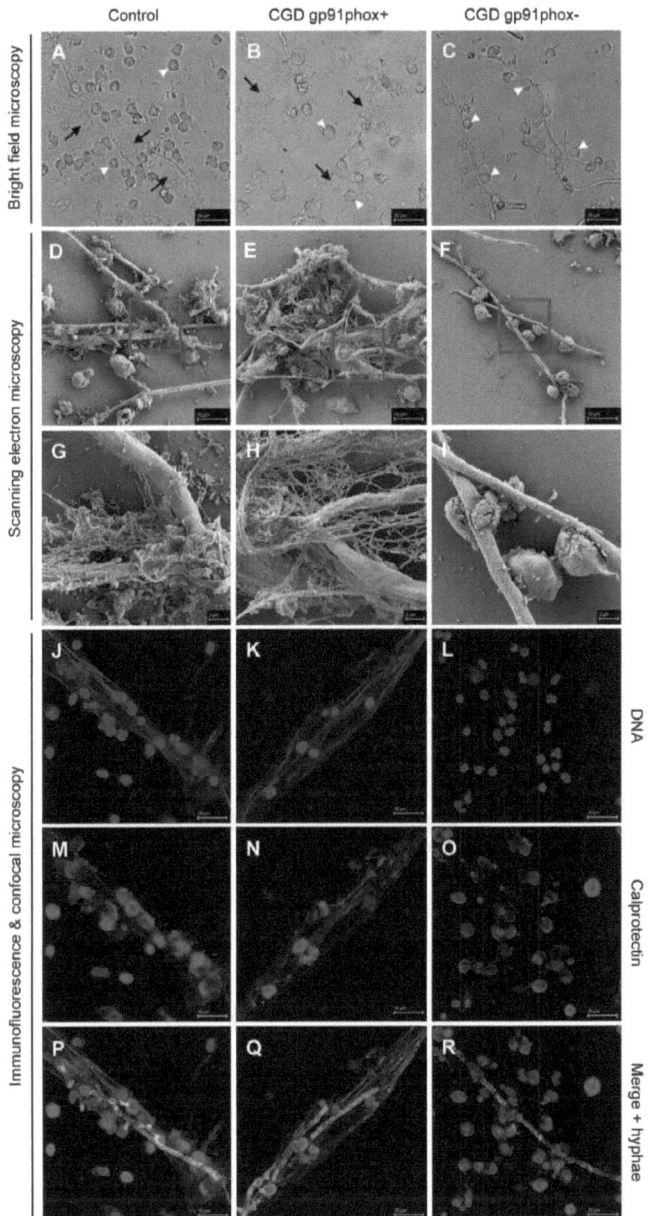

FIG 5. *A. nidulans* induced NETs contain calprotectin. Control and CGD sorted neutrophils were infected with hyphae and NET-formation was determined by bright field microscopy (A-C), SEM (D-I), and immunofluorescence and confocal microscopy (J-R). NET-formation was determined by DAPI (blue), calprotectin by staining with antibodies against S100A8/A9 (red) and hyphae with antibodies against *Aspergillus* soluble proteins (green).

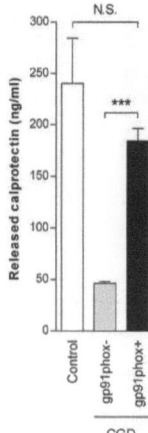

FIG 6. Calprotectin release by gp91[phox+] and gp91[phox-] neutrophils. Calprotectin release was quantified by ELISA after neutrophil activation by PMA inducing NET-formation. NETs were digested by MNase/DNase-1 to quantify NET-associated calprotectin. Significance was assessed by Student's t-test: N.S. = not significant ($P > 0.05$), ***$P < 0.001$.

Discussion

Recent work suggests NET contribution to elimination of fungal infections in healthy subjects since hyphae are too large to be phagocytosed [23-24]. CGD patients are unable to make NETs [5, 11] and are susceptible to therapy-refractory infection with *Aspergillus spp* [10]. Especially infections with *A. nidulans* are a major threat to these patients, due to higher virulence and frequent resistance to antifungal treatment, as compared to *A. fumigatus* [7]. Here we show that genetic complementation of NADPH oxidase by human CGD GT restored the ability of neutrophils to release NETs and NET-associated calprotectin, which is responsible for growth inhibition of *A. nidulans* by Zn^{2+} sequestration. A threshold of 27.8% of O_2^- production per individual gp91[phox+] neutrophil and a chimerism of 22.8% of oxidase-positive cells was enough for sufficient NET-formation *in vitro* and clinical clearance of multifocal therapy-refractory *A. nidulans* lung infection in a patient with CGD *in vivo*.

The S100 Ca^{2+}-binding protein calprotectin is a heteroduplex of subunits S100A8 and S100A9 which is expressed by granulocytes, monocytes, and early differentiation stages of macrophages [25]. It represents about 40 % of neutrophil cytoplasm [26], accumulates at high concentrations (1-20 mg/ml) in abscess-fluid supernatants [27] and has been shown to bind Zn^{2+} *in vitro* [28]. Its growth inhibitory effect on fungi is most likely mediated by metal-chelation and is reversible by micro-molar quantities of Zn^{2+} [18, 27-28]. We set out to study the role of calprotectin in innate defense against CGD relevant *A. nidulans*. Growth of *A. nidulans* was strongly inhibited when coincubated with both purified calprotectin subunits, which is in accordance with a report by Sohnle et al [17] and with crystal structure [29] as well as mass spectrometric analyses [30] suggesting that the Ca^{2+}-dependent formation of S100A8/A9 heterotetramers would lead to enhanced Zn^{2+}-binding capacity, whilst none of the homodimers should display significant affinity for Zn^{2+} [29].

Our study demonstrates the major role of NET-associated calprotectin in suppressing growth of *A. nidulans* conidia and hyphae, and shows presence of calprotectin in NETs induced by *A. nidulans* hyphae. *A. fumigatus* has been reported recently to trigger NET release [31-32], resulting in growth inhibition of *A. fumigatus* which could be prevented by Zn^{2+} [32]. However, the mechanism of inhibition was not clearly shown in these reports. Here we show that blocking of calprotectin by specific antibodies and the absence of calprotectin in NETs from calprotectin-deficient mouse neutrophils completely abrogated the inhibition of *A. nidulans* conidia and hyphae growth. This is a more precise analysis, since NETs also contain other Zn^{2+} binding proteins, such as S100A12 [33] which is involved in anti-parasite responses.

According to our findings calprotectin has to be presented extracellularly for antifungal activity against *A. nidulans*. To the best of our knowledge, leakage of a cytoplasmic protein into phagolysosomes has not been described. We previously showed that neutrophil degranulation stimulated by the bacterial peptide fMLP does not induce secretion of calprotectin, indicating non-vesicular localization [4]. However, we cannot entirely exclude that minor amounts of calprotectin are found in phagolysosomes and contribute to the slight inhibitory effect we reported on CGD neutrophils infected with *A. nidulans* conidia.

We show that presence of functional NADPH oxidase in human neutrophils is an absolute requirement for *A. nidulans* growth inhibition by NET-associated calprotectin, as only gp91^{phox+} but not gp91^{phox-} neutrophils made NETs containing calprotectin and inhibited *A. nidulans* growth efficiently. GT of human CGD restored the amount of released calprotectin during NET-formation to wild-type levels and NET-associated calprotectin co-localized with *A. nidulans* hyphae. Analysis of gp91^{phox-} neutrophils enabled us to establish the role of NADPH oxidase for the release of calprotectin during NET formation, which is an advantage over chemical inhibition of NADPH oxidase using diphenylene iodonium chloride (DPI). We propose that NET formation is an NADPH oxidase-dependent mechanism by which neutrophils secrete calprotectin into the extracellular compartment upon infection with *Aspergillus* hyphae, resulting in both NET-bound and free calprotectin. NET-associated calprotectin is concentrated with trapped *Aspergillus* for antifungal function. Unbound calprotectin may additionally serve for distinct functions, such as chemotaxis or inflammation.

The fact that gp91^{phox-} neutrophils also partially inhibited conidia germination, suggests that NADPH oxidase-independent mechanisms linked to phagocytosis, such as activity of the granule serine proteases NE and cathepsin G [34-36] (while the mechanism has been recently object of controversy [37-39]), might contribute to early defense against *Aspergillus*. The need of NE for NET formation [14] could explain the susceptibility to infection of NE deficient mice described by Revees et al [35] as an indirect action. Other mechanisms such as the tryptophan catabolizing enzyme indoleamine-2,3-dioxygenase (IDO), considered critical for regulating immune responses and suppression of inflammation in mice [40], do not play a role in human CGD as we previously reported [41-42].

Fungi have high Zn^{2+} requirement (10^{-7}-10^{-5} M) for growth [43] with a low minimum inhibitory concentration of calprotectin [44]. Due to the lack of a well-developed extracellular Zn^{2+}-scavenging system, *A. fumigatus* is extremely sensitive to Zn^{2+} deprivation [45]. Zn^{2+} requirement for *A. nidulans* in our study was in the same range as described for other fungi. In contrast to *S. aureus* which is more sensitive to calprotectin-mediated Mn^{2+} deprivation [19], we found that *A. nidulans* could only resume growth after Zn^{2+} supplementation, but not after Mn^{2+}, Fe^{2+} or Cu^{2+} addition. This supports the central role of Zn^{2+} for *A. nidulans* growth.

Interestingly, *A. nidulans* hyphae were not killed by NETs at low calprotectin concentrations *in vitro*, although hyphal structures seemed to be severely deranged by NETs. This corresponded to the range of fungistatic calprotectin concentrations described for *Candida* [44]. *A. nidulans* growth was restored after addition of Zn^{2+}, even after a month of inhibition by NETs. As much higher calprotectin concentrations have been observed in human abscesses [27], *Aspergillus* hyphae might still be killed by NET-associated calprotectin *in vivo*. Killing of *Candida spp.* or *Cryptococcus neoformans* was demonstrated at calprotectin concentrations of 3-5 µg/ml [44]. We show for the first time that calprotectin concentrations ≥ 16µg/ml lead to irreversible growth arrest of *A. nidulans* conidia and hyphae, most likely by Zn^{2+} starvation. In contrast to our findings a previous publication reported only minor antifungal activity of NETs against *A. fumigatus* [31], probably due to a shorter incubation period. Alternatively, complete *in vivo* killing might also be achieved by other antimicrobial peptides [46] or by so far unknown mechanisms: in addition to their antimicrobial properties, S100A8, S100A9, and S100A8/A9 are chemotactic for neutrophils and monocytes [47-49]. Notably, for this function calprotectin needs to be extracellular, and thus *Aspergillus*-induced NET formation supports it as well.

Here we establish that calprotectin released in NETs displays concentration-dependent reversible and irreversible growth inhibitory activity against virulent *A. nidulans*. NETs may be involved in disarming *A. nidulans* and may prevent further spreading by providing high local concentrations of calprotectin. Definitive *in vivo* proof in humans is obviously ethically impossible. In conclusion, we show that calprotectin is a critical factor in the innate immune defense of human neutrophils to *Aspergillus* infection and adds to the concept of metal chelation as a strategy for inhibiting microbial growth at the sites of infection. The severe immunodeficient phenotype and the high susceptibility to *Aspergillus* infection of CGD patients might be linked to absence of NETs, and restoration of NADPH oxidase function and NET-formation by GT leads to rapid cure of refractory invasive aspergillosis in X-linked CGD.

Clinical implications

Gene therapy for CGD restores NET-formation and calprotectin release by neutrophils, leading to efficient *A. nidulans* inhibition. Calprotectin is a critical factor in human innate immune defense against *Aspergillus* infection.

Acknowledgments

The authors wish to thank the patient and his family for their trust. We are indebted to the medical and nursing staff of the bone marrow transplantation unit of University Children's Hospital Zurich. We would like to thank Reinhard Seger for his support and helpful discussions, and Manuel Grez and Klaus Kühlke for developing and providing the SF71gp91^{phox} vector, respectively. We are grateful to Klaus Marquardt for electron microscopy, to Andres Kaech for help with confocal microscopy, and to Alex Imhof for isolating *A. nidulans* conidia. Furthermore we wish to acknowledge Thomas Vogl and Johannes Roth for supplying *S100A9* knockout mice.

References

1. Brinkmann V, Reichard U, Goosmann C, Fauler B, Uhlemann Y, Weiss DS, et al. Neutrophil extracellular traps kill bacteria. Science 2004; 303:1532-5.
2. Guimaraes-Costa AB, Nascimento MT, Froment GS, Soares RP, Morgado FN, Conceicao-Silva F, et al. Leishmania amazonensis promastigotes induce and are killed by neutrophil extracellular traps. Proc Natl Acad Sci U S A 2009; 106:6748-53.
3. Urban CF, Reichard U, Brinkmann V, Zychlinsky A. Neutrophil extracellular traps capture and kill Candida albicans yeast and hyphal forms. Cell Microbiol 2006; 8:668-76.
4. Urban CF, Ermert D, Schmid M, Abu-Abed U, Goosmann C, Nacken W, et al. Neutrophil extracellular traps contain calprotectin, a cytosolic protein complex involved in host defense against Candida albicans. PLoS Pathog 2009; 5:e1000639.
5. Fuchs TA, Abed U, Goosmann C, Hurwitz R, Schulze I, Wahn V, et al. Novel cell death program leads to neutrophil extracellular traps. J Cell Biol 2007; 176:231-41.
6. Hsu K, Champaiboon C, Guenther BD, Sorenson BS, Khammanivong A, Ross KF, et al. ANTI-INFECTIVE PROTECTIVE PROPERTIES OF S100 CALGRANULINS. Antiinflamm Antiallergy Agents Med Chem 2009; 8:290-305.
7. Segal BH, DeCarlo ES, Kwon-Chung KJ, Malech HL, Gallin JI, Holland SM. Aspergillus nidulans infection in chronic granulomatous disease. Medicine (Baltimore) 1998; 77:345-54.
8. Seger RA. Modern management of chronic granulomatous disease. Br J Haematol 2008; 140:255-66.
9. Winkelstein JA, Marino MC, Johnston RB, Jr., Boyle J, Curnutte J, Gallin JI, et al. Chronic granulomatous disease. Report on a national registry of 368 patients. Medicine (Baltimore) 2000; 79:155-69.
10. Segal BH, Romani LR. Invasive aspergillosis in chronic granulomatous disease. Med Mycol 2009; 47 Suppl 1:S282-90.
11. Bianchi M, Hakkim A, Brinkmann V, Siler U, Seger RA, Zychlinsky A, et al. Restoration of NET formation by gene therapy in CGD controls aspergillosis. Blood 2009;144:2619-22.
12. Ermert D, Urban CF, Laube B, Goosmann C, Zychlinsky A, Brinkmann V. Mouse neutrophil extracellular traps in microbial infections. J Innate Immun 2009; 1:181-93.
13. Metzler KD, Fuchs TA, Nauseef WM, Reumaux D, Roesler J, Schulze I, et al. Myeloperoxidase is required for neutrophil extracellular trap formation: implications for innate immunity. Blood 2010;117:953-59.

14. Papayannopoulos V, Metzler KD, Hakkim A, Zychlinsky A. Neutrophil elastase and myeloperoxidase regulate the formation of neutrophil extracellular traps. J Cell Biol 2010; 191:677-91.
15. Bonnett CR, Cornish EJ, Harmsen AG, Burritt JB. Early neutrophil recruitment and aggregation in the murine lung inhibit germination of Aspergillus fumigatus Conidia. Infect Immun 2006; 74:6528-39.
16. Mayo LA, Curnutte JT. Kinetic microplate assay for superoxide production by neutrophils and other phagocytic cells. Methods Enzymol 1990; 186:567-75.
17. Sohnle PG, Hunter MJ, Hahn B, Chazin WJ. Zinc-reversible antimicrobial activity of recombinant calprotectin (migration inhibitory factor-related proteins 8 and 14). J Infect Dis 2000; 182:1272-5.
18. Murthy AR, Lehrer RI, Harwig SS, Miyasaki KT. In vitro candidastatic properties of the human neutrophil calprotectin complex. J Immunol 1993; 151:6291-301.
19. Corbin BD, Seeley EH, Raab A, Feldmann J, Miller MR, Torres VJ, et al. Metal chelation and inhibition of bacterial growth in tissue abscesses. Science 2008; 319:962-5.
20. Sohnle PG, Hahn BL, Santhanagopalan V. Inhibition of Candida albicans growth by calprotectin in the absence of direct contact with the organisms. J Infect Dis 1996; 174:1369-72.
21. Behnsen J, Narang P, Hasenberg M, Gunzer F, Bilitewski U, Klippel N, et al. Environmental dimensionality controls the interaction of phagocytes with the pathogenic fungi Aspergillus fumigatus and Candida albicans. PLoS Pathog 2007; 3:e13.
22. Nacken W, Kerkhoff C. The hetero-oligomeric complex of the S100A8/S100A9 protein is extremely protease resistant. FEBS Lett 2007; 581:5127-30.
23. Latge JP. Aspergillus fumigatus and aspergillosis. Clin Microbiol Rev 1999; 12:310-50.
24. Mizgerd JP. Acute lower respiratory tract infection. N Engl J Med 2008; 358:716-27.
25. Ehrchen JM, Sunderkotter C, Foell D, Vogl T, Roth J. The endogenous Toll-like receptor 4 agonist S100A8/S100A9 (calprotectin) as innate amplifier of infection, autoimmunity, and cancer. J Leukoc Biol 2009; 86:557-66.
26. Edgeworth J, Gorman M, Bennett R, Freemont P, Hogg N. Identification of p8,14 as a highly abundant heterodimeric calcium binding protein complex of myeloid cells. J Biol Chem 1991; 266:7706-13.
27. Clohessy PA, Golden BE. Calprotectin-mediated zinc chelation as a biostatic mechanism in host defence. Scand J Immunol 1995; 42:551-6.

28. Sohnle PG, Collins-Lech C, Wiessner JH. The zinc-reversible antimicrobial activity of neutrophil lysates and abscess fluid supernatants. J Infect Dis 1991; 164:137-42.
29. Korndorfer IP, Brueckner F, Skerra A. The crystal structure of the human (S100A8/S100A9)2 heterotetramer, calprotectin, illustrates how conformational changes of interacting alpha-helices can determine specific association of two EF-hand proteins. J Mol Biol 2007; 370:887-98.
30. Vogl T, Leukert N, Barczyk K, Strupat K, Roth J. Biophysical characterization of S100A8 and S100A9 in the absence and presence of bivalent cations. Biochim Biophys Acta 2006; 1763:1298-306.
31. Bruns S, Kniemeyer O, Hasenberg M, Aimanianda V, Nietzsche S, Thywissen A, et al. Production of extracellular traps against Aspergillus fumigatus in vitro and in infected lung tissue is dependent on invading neutrophils and influenced by hydrophobin RodA. PLoS Pathog 2010; 6:e1000873.
32. McCormick A, Heesemann L, Wagener J, Marcos V, Hartl D, Loeffler J, et al. NETs formed by human neutrophils inhibit growth of the pathogenic mold Aspergillus fumigatus. Microbes Infect 2010;12:928-36.
33. Moroz OV, Blagova EV, Wilkinson AJ, Wilson KS, Bronstein IB. The crystal structures of human S100A12 in apo form and in complex with zinc: new insights into S100A12 oligomerisation. J Mol Biol 2009; 391:536-51.
34. Decleva E, Menegazzi R, Busetto S, Patriarca P, Dri P. Common methodology is inadequate for studies on the microbicidal activity of neutrophils. J Leukoc Biol 2006; 79:87-94.
35. Reeves EP, Lu H, Jacobs HL, Messina CG, Bolsover S, Gabella G, et al. Killing activity of neutrophils is mediated through activation of proteases by K+ flux. Nature 2002; 416:291-7.
36. Tkalcevic J, Novelli M, Phylactides M, Iredale JP, Segal AW, Roes J. Impaired immunity and enhanced resistance to endotoxin in the absence of neutrophil elastase and cathepsin G. Immunity 2000; 12:201-10.
37. Ahluwalia J, Tinker A, Clapp LH, Duchen MR, Abramov AY, Pope S, et al. The large-conductance Ca2+-activated K+ channel is essential for innate immunity. Nature 2004; 427:853-8.
38. Essin K, Salanova B, Kettritz R, Sausbier M, Luft FC, Kraus D, et al. Large-conductance calcium-activated potassium channel activity is absent in human and mouse neutrophils and is not required for innate immunity. Am J Physiol Cell Physiol 2007; 293:C45-54.

39. Femling JK, Cherny VV, Morgan D, Rada B, Davis AP, Czirjak G, et al. The antibacterial activity of human neutrophils and eosinophils requires proton channels but not BK channels. J Gen Physiol 2006; 127:659-72.
40. Romani L, Fallarino F, De Luca A, Montagnoli C, D'Angelo C, Zelante T, et al. Defective tryptophan catabolism underlies inflammation in mouse chronic granulomatous disease. Nature 2008; 451:211-5.
41. Hakkim A, Hurwitz R, Bianchi M, Brinkmann V, Siler U, Seger R, et al. Protecting against Aspergillus infection in CGD Blood 2009; 114:3498.
42. Jurgens B, Fuchs D, Reichenbach J, Heitger A. Intact indoleamine 2,3-dioxygenase activity in human chronic granulomatous disease. Clin Immunol 2010;137:1-4.
43. Sugarman B. Zinc and infection. Rev Infect Dis 1983; 5:137-47.
44. Steinbakk M, Naess-Andresen CF, Lingaas E, Dale I, Brandtzaeg P, Fagerhol MK. Antimicrobial actions of calcium binding leucocyte L1 protein, calprotectin. Lancet 1990; 336:763-5.
45. Lulloff SJ, Hahn BL, Sohnle PG. Fungal susceptibility to zinc deprivation. J Lab Clin Med 2004; 144:208-14.
46. Levitz SM, Selsted ME, Ganz T, Lehrer RI, Diamond RD. In vitro killing of spores and hyphae of Aspergillus fumigatus and Rhizopus oryzae by rabbit neutrophil cationic peptides and bronchoalveolar macrophages. J Infect Dis 1986; 154:483-9.
47. Ryckman C, Vandal K, Rouleau P, Talbot M, Tessier PA. Proinflammatory activities of S100: proteins S100A8, S100A9, and S100A8/A9 induce neutrophil chemotaxis and adhesion. J Immunol 2003; 170:3233-42.
48. Vandal K, Rouleau P, Boivin A, Ryckman C, Talbot M, Tessier PA. Blockade of S100A8 and S100A9 suppresses neutrophil migration in response to lipopolysaccharide. J Immunol 2003; 171:2602-9.
49. Raquil MA, Anceriz N, Rouleau P, Tessier PA. Blockade of antimicrobial proteins S100A8 and S100A9 inhibits phagocyte migration to the alveoli in streptococcal pneumonia. J Immunol 2008; 180:3366-74.

Online repository

Methods

Patient description and gene therapy (GT)

We treated an 8 7/12 years old boy with X-linked gp91^{phox}-deficient CGD and multifocal therapy refractory *A. nidulans* lung infection with a monocistronic LTR-driven gammaretroviral SF71gp91^{phox} vector (details in [1]) under a protocol approved by the ethics review board of the University Children's Hospital Zurich and the Swiss Expert Committee for Bio-Safety after written informed consent of the parents. Collection of CD34$^+$ cells, transduction, conditioning with low-dose busulfan IV (8.8 mg/KG), and clinical follow up were performed as described [2]. The *A. nidulans* lung infection was completely cleared six weeks after GT. Therapy/prophylaxis with oral voriconazole was continued throughout and has not yet been tapered 2.6 years after GT. To avoid interactions, voriconazole was stopped four days prior to *Aspergillus* experiments and continued thereafter. At time of analysis the patient was free of any infection, 22.8% of neutrophils expressed gp91^{phox}, with an O_2^- production of 27.8% per gp91^{phox} expressing cell (relative to healthy controls) [3]. Clinical and molecular follow-up of GT in this patient will be described elsewhere.

Aspergillus nidulans strain

The *A. nidulans* strain used was isolated from bronchoalveolar lavage fluid of the patient; conidia were grown and collected as described [4]. For all experiments conidia and hyphae were grown in serum free RPMI medium (phenol red-free) supplemented with 10 mM Hepes. If not stated otherwise, fungal growth was quantified in all assays with the tetrazolium dye 2,3-bis(2-Methoxy-4-nitro-5-sulfophenyl)-2H-tetrazolium-5-carboxanilide (XTT, Invitrogen) as described [5]. This method, unlike colony forming unit (CFU) enumeration, does not require dispersion of the hyphae.

Isolation of human and murine neutrophils

Human neutrophils were isolated from peripheral blood of healthy donors or the CGD patient using dextran-Ficoll (GE Healthcare) [6] for all NET assays with *A. nidulans* and for autoMACS sorting, whereas Percoll (GE Healthcare) density gradient separation [7] was used for NET quantification assays. For all experiments, neutrophils were resuspended in serum free RPMI medium (phenol red-free) supplemented with 10 mM Hepes and used within 1 hour after isolation.

Murine neutrophils were isolated from *S100A9$^{-/-}$* mice backcrossed 10 times into C57 BL/6. These mice are deficient in both calprotectin subunits S100A8 and S100A9 protein [8]. Mice were bred in our animal facility according to regulations of the Jordbruksverket Sweden (Dnr A29-69). Mature murine neutrophils were isolated from bone marrow as previously described [9]. Briefly, bone marrow cells from tibia and femur were singularized using a 70 μm cell strainer and separated by centrifugation for 30 min at 1,500 g on a discontinuous Percoll gradient with 52 % (v/v), 69 % (v/v), and 78 % (v/v). Neutrophils harvested from the distinct layer between 69 % and 78 % were resuspended in HBSS without Ca^{2+} and Mg^{2+}.

Sorting of CGD gp91^{phox+} and gp91^{phox-} neutrophils

Control and patient neutrophils were first stained 30 min at room temperature with 10 µg/ml mouse α- gp91^{phox}-FITC antibody (clone 7D5, MBL), then 30 min at 4°C with anti-FITC MicroBeads (Miltenyi Biotec) for magnetic separation by autoMACS (Miltenyi Biotec), according to the manufacturer's instructions. Purity of resulting gp91^{phox}-negative (gp91^{phox-}) and -positive (gp91^{phox+}) populations was directly measured by the fluorescent-activated cell sorter (FACS).

A. nidulans growth inhibition by purified calprotectin

In a 96-well plate, 10^4 A. nidulans conidia or hyphae obtained from culture of 10^4 conidia were incubated 24 h at 37°C with 5 µg/ml recombinant S100A8 and/or 5 µg/ml recombinant S100A9 (ProtEra) ± 1 µM $ZnSO_4$ or 15 µg/ml rabbit polyclonal α-S100A9 antibody (H00006280-D01P, Abnova) vs. unspecific control antibody (Sigma-Aldrich) [10]. This α-S100A9 antibody recognizes the whole S100A9 subunit alone or the S100A9 subunit complexed with S100A8. Fungal growth is expressed as percentage of control values (A. nidulans conidia or hyphae incubated in media without neutrophils ± $ZnSO_4$, S100A8/A9 or antibodies).

A. nidulans growth inhibition by NETs

In a 96-well plate, 10^5 control, CGD gp91^{phox-} and gp91^{phox+} sorted neutrophils were activated with 40 nM PMA (Sigma-Aldrich) at 37°C and 5% CO_2 for 4 h, until NET formation was complete. 0 to 2 µM $ZnSO_4$ (Sigma-Aldrich) and A. nidulans conidia or hyphae (hyphae previously grown O/N in a 24-well plate, collected by pipetting and scraping the wells with a cell scraper) with a multiplicity of infection (MOI) A. nidulans/neutrophils of 0.1 were then added, in a final volume of 160 µl. Afterwards the plates were centrifuged 5 min at 400 g and incubated 24 h at 37°C, allowing conidia germination and hyphal outgrowth. Fungal growth is expressed as percentage of control values (A. nidulans conidia or hyphae incubated in media ± $ZnSO_4$).

A. nidulans growth inhibition by NET-extracts

NET formation was induced in $8x10^5$ control, CGD gp91^{phox-} and gp91^{phox+} neutrophils by activation with 40 nM PMA at 37°C and 5% CO_2 for 5 h in a 24-well plate (500 µl/well; 4 wells each; $3.2x10^6$ cells total). For maximal DNA fragmentation NETs were then digested for 30 min with 5 U/ml micrococcal nuclease (MNase, Worthington Biochemical) + 1 U/ml deoxyribonuclease (DNase)-1 (Sigma-Aldrich), and pooled supernatant concentrated ~ 10-fold by centrifugation on filter columns with 3 kDa cut-off (Amicon, Millipore), following the manufacturer's instructions. In a 96-well plate, 10^4 A. nidulans conidia or hyphae obtained from culture of 6×10^4 *A. nidulans* conidia were then incubated 24 h at 37°C with the concentrated NET-extracts (diluted 1/3) with or without 15 µg/ml α-S100A9 or unspecific control antibody in 100 µl final volume. Fungal growth is expressed as percentage of control values (*A. nidulans* conidia or hyphae incubated in media ± antibodies).

A. nidulans growth inhibition by murine $S100A9^{-/-}$ NETs

In a 24-well plate, 5×10^5 mouse neutrophils in 500 µl RPMI with 1% (v/v) mouse serum were stimulated with 100 nM PMA for 20 h at 37°C with 5% CO_2 to induce NET formation. The supernatant was discarded; NETs were washed once with RPMI and incubated with 500 µl RPMI containing *A. nidulans* conidia or hyphae at MOI 0.1 and 0.01, respectively. Afterwards plates were centrifuged for 5 min at 300 g and incubated 20 h at 37°C. Fungal growth is expressed as percentage of control values (*A. nidulans* conidia or hyphae incubated in media).

Fungistatic and fungicidal effect of calprotectin

In a 96-well plate, 10^4 *A. nidulans* conidia or hyphae obtained from culture of 10^4 conidia were incubated on previously formed NETs (from 10^5 neutrophils activated with 40 nM PMA at 37°C and 5% CO_2 for 4 h) or with 0.5-128 µg/ml purified S100A8/A9. After 4 days to 4 weeks, RPMI or $ZnSO_4$ (final concentration 1 µM for NETs and 50 µM for purified S100A8/A9) were added to the wells and incubated O/N. Growth of *A. nidulans* conidia or hyphae was assessed using an inverted light microscope (Leica DM IL HC Fluo) coupled to a CCD camera (Leica DFC 480 R2).

Statistical analysis

Two-tailed Student's *t*-test was used for analysis of two groups. Differences were considered statistically significant when $p < 0.05$. All statistical tests were performed using GraphPad Prism.

References

E1. Bianchi M, Hakkim A, Brinkmann V, Siler U, Seger RA, Zychlinsky A, et al. Restoration of NET formation by gene therapy in CGD controls aspergillosis. Blood 2009;144:2619-22.

E2. Ott MG, Schmidt M, Schwarzwaelder K, Stein S, Siler U, Koehl U, et al. Correction of X-linked chronic granulomatous disease by gene therapy, augmented by insertional activation of MDS1-EVI1, PRDM16 or SETBP1. Nat Med 2006; 12:401-9.

E3. Mayo LA, Curnutte JT. Kinetic microplate assay for superoxide production by neutrophils and other phagocytic cells. Methods Enzymol 1990; 186:567-75.

E4. Bonnett CR, Cornish EJ, Harmsen AG, Burritt JB. Early neutrophil recruitment and aggregation in the murine lung inhibit germination of Aspergillus fumigatus Conidia. Infect Immun 2006; 74:6528-39.

E5. Meshulam T, Levitz SM, Christin L, Diamond RD. A simplified new assay for assessment of fungal cell damage with the tetrazolium dye, (2,3)-bis-(2-methoxy-4-nitro-5-sulphenyl)-(2H)-tetrazolium-5-carboxanilide (XTT). J Infect Dis 1995; 172:1153-6.

E6. Weiss J, Kao L, Victor M, Elsbach P. Oxygen-independent intracellular and oxygen-dependent extracellular killing of Escherichia coli S15 by human polymorphonuclear leukocytes. J Clin Invest 1985; 76:206-12.

E7. Aga E, Katschinski DM, van Zandbergen G, Laufs H, Hansen B, Muller K, et al. Inhibition of the spontaneous apoptosis of neutrophil granulocytes by the intracellular parasite Leishmania major. J Immunol 2002; 169:898-905.

E8. Manitz MP, Horst B, Seeliger S, Strey A, Skryabin BV, Gunzer M, et al. Loss of S100A9 (MRP14) results in reduced interleukin-8-induced CD11b surface expression, a polarized microfilament system, and diminished responsiveness to chemoattractants in vitro. Mol Cell Biol 2003; 23:1034-43.

E9. Ermert D, Urban CF, Laube B, Goosmann C, Zychlinsky A, Brinkmann V. Mouse neutrophil extracellular traps in microbial infections. J Innate Immun 2009; 1:181-93.

E10. Sohnle PG, Hunter MJ, Hahn B, Chazin WJ. Zinc-reversible antimicrobial activity of recombinant calprotectin (migration inhibitory factor-related proteins 8 and 14). J Infect Dis 2000; 182:1272-5.

Figure E1

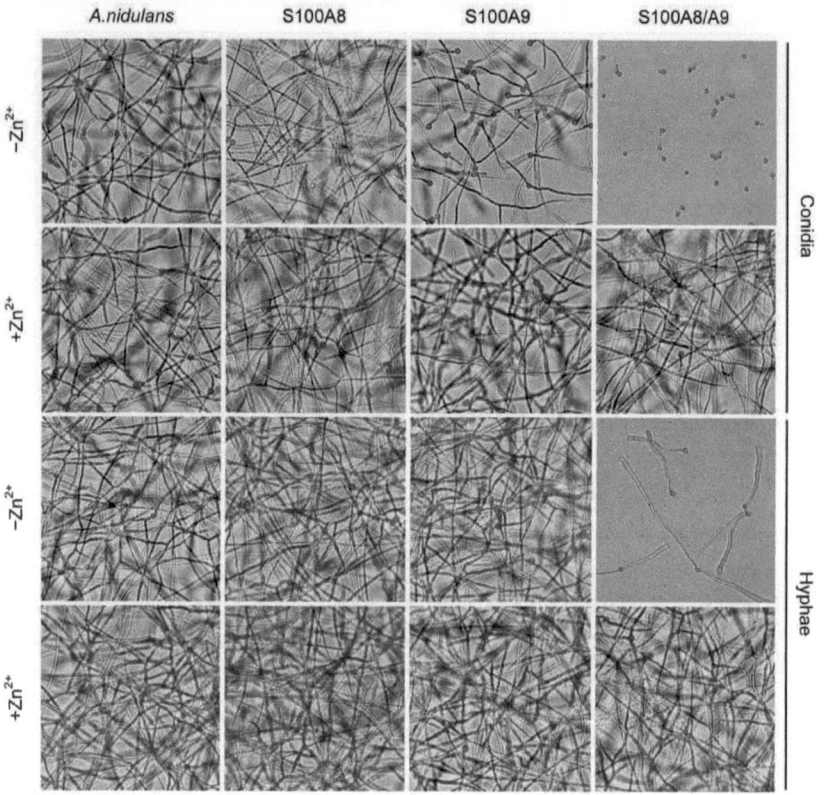

FIG E1. Inhibition of *A. nidulans* growth by purified calprotectin is abolished by Zn^{2+} addition. Incubation with the S100A8 (5 µg/ml) subunit did not prevent *A. nidulans* conidia or hyphae growth, whereas incubation with S100A9 (5 µg/ml) partially inhibited conidia germination. Coincubation with both subunits strongly inhibited conidia germination and hyphae growth. In all cases, antifungal activity was completely abolished in presence of 1 µM Zn^{2+}.

Figure E2

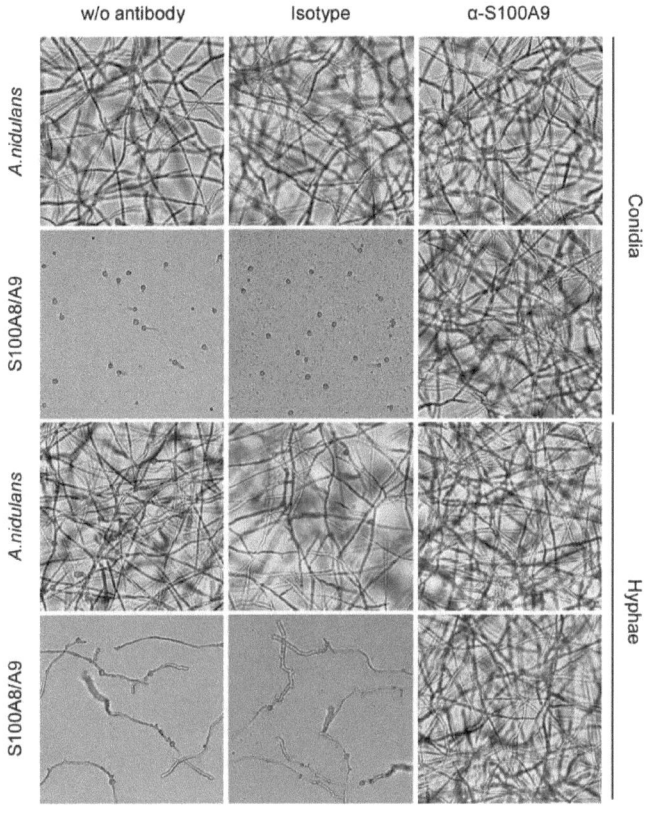

FIG E2. Inhibition of *A. nidulans* growth by purified calprotectin is abolished by α-S100A9 antibody. 15 µg/ml polyclonal α-S100A9 antibody fully inhibited antifungal activity of the S100A8/A9 calprotectin complex on conidia or hyphae. As control, S100A8/A9 was incubated with an unspecific control antibody, showing no effect on antifungal activity of the S100A8/A9 calprotectin complex. w/o, without.

Figure E3

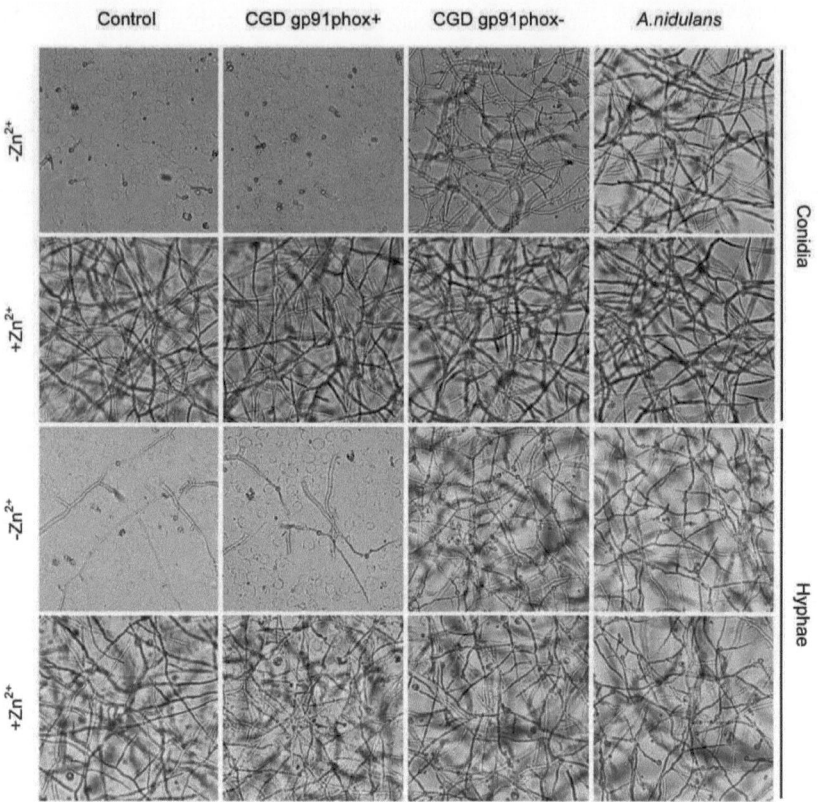

FIG E3. Calprotectin-dependent antifungal activity of CGD gp91^{phox+} neutrophils is abolished by Zn^{2+} addition. Control and CGD sorted neutrophils were stimulated with PMA for NET formation and incubated overnight with *A. nidulans* conidia or hyphae in presence or absence of 1 µM Zn^{2+}. gp91^{phox+} neutrophils showed strong antifungal activity against conidia or hyphae, comparable to control. gp91^{phox-} neutrophils were inefficient in controlling fungal growth, with only partial inhibition of conidia germination. In all cases, antifungal activity was completely abolished in presence of 1 µM Zn^{2+}.

Figure E4

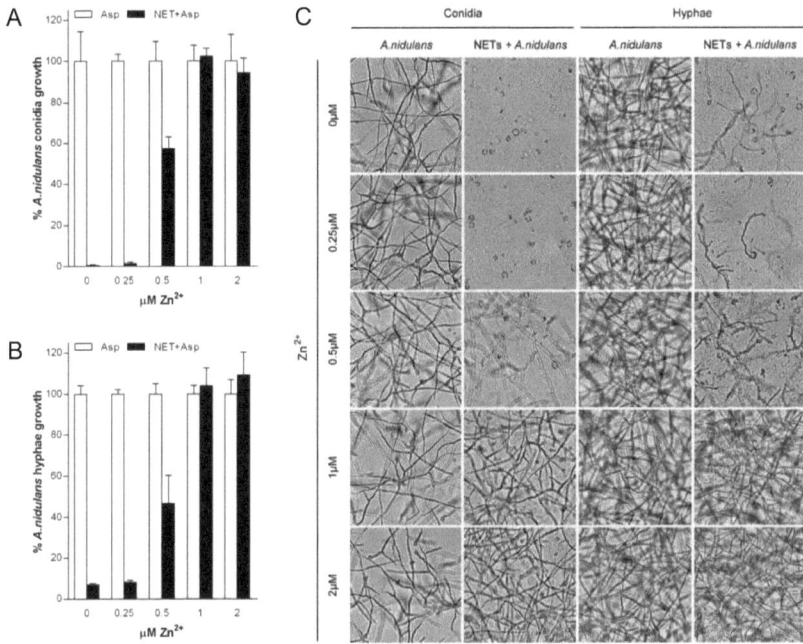

FIG E4. NET-inhibition of *A. nidulans* growth is abolished by Zn^{2+} addition. Human neutrophils were stimulated with PMA for NET formation and subsequently infected overnight with *A. nidulans* conidia or hyphae ± Zn^{2+}. Increasing concentrations of Zn^{2+} abolished the antifungal activity of NETs against conidia (A) or hyphae (B). (C) Representative pictures of data shown in (A-B).

Figure E5

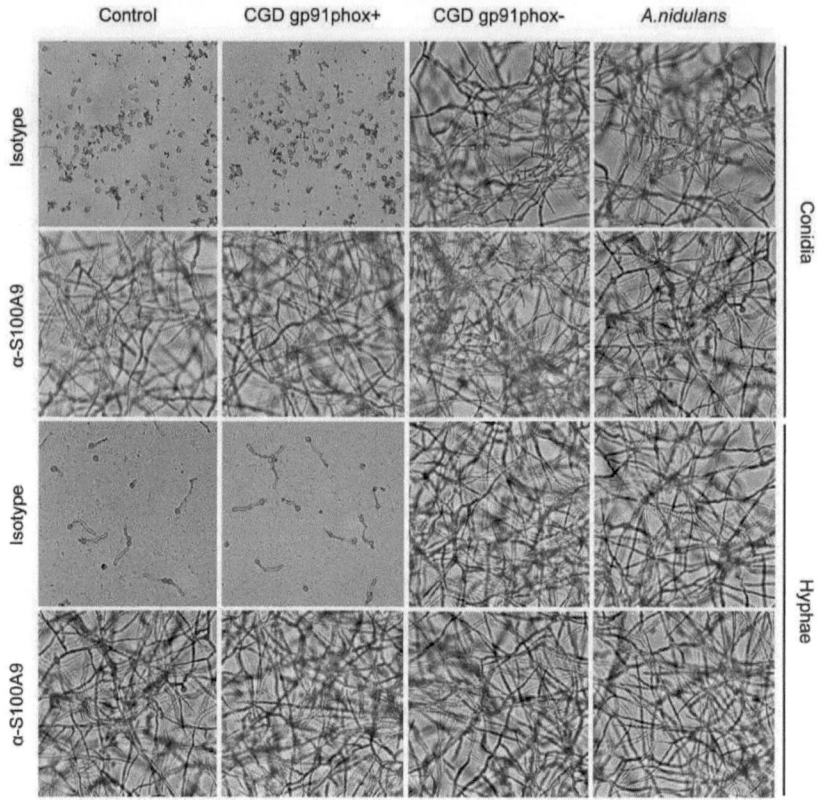

FIG E5. Calprotectin-dependent antifungal activity of CGD gp91^{phox+} neutrophils is abolished by α-S100A9 antibody. Control and CGD sorted neutrophils were stimulated with PMA for NET formation; NET supernatant was digested by DNase/MNase, concentrated and incubated with *A. nidulans* conidia or hyphae after previous incubation with 15 µg/ml polyclonal α-S100A9 or unspecific control antibody. NET extracts from control and gp91^{phox+} neutrophils showed strong antifungal activity against conidia or hyphae. Supernatant from gp91^{phox-} neutrophils had no inhibitory activity on fungal growth. In all cases, fungal growth was completely restored by addition of α-S100A9 antibody proving that the restored antifungal activity of gp91^{phox+} neutrophils after GT was calprotectin dependent.

Letter sent to *JACI* regarding the paper: Bianchi M, et al. Restoration of anti-*Aspergillus* defense by neutrophil extracellular traps in human chronic granulomatous disease after gene therapy is calprotectin-dependent. *J Allergy Clin Immunol* 2011; online June 20

CORRESPONDENCE

Redundant ability of phagocytes to kill *Aspergillus* species

Joachim Roesler MD, PhD, and Angela Rösen-Wolff MD, PhD

Dept. of Pediatrics, University Hospital *Carl Gustav Carus*, Dresden, Germany

Corresponding Author:
Joachim Roesler MD, PhD
Dept. of Pediatrics
University Hospital *Carl Gustav Carus*
Fetscherstr. 74, 01307 Dresden, Germany
Tel.: +49 351 458 6870, Fax: +49 351 458 4384
e-mail: Roeslerj@rcs.urz.tu-dresden.de

CORRESPONDENCE

Redundant ability of phagocytes to kill *Aspergillus* species

To the Editor:

Recently, Bianchi et al[1] published an excellent article demonstrating the efficacy of neutrophil extracellular traps (NETs) in killing *Aspergillus nidulans*. The authors show that this killing is mediated by calprotectin associated with NETs. Neutrophils cannot phagocytose long *Aspergillus* hyphae, but they can form NETs along such hyphae and thereby effectively destroy these fungal structures. In contrast, neutrophils from patients with chronic granulomatous disease (CGD) are unable to form NETs because NET formation depends on the production of reactive oxygen species (ROS). ROS are normally produced by the multienzyme complex nicotinamide adenine dinucleotide phosphate (NADPH) oxidase, which is defective in CGD. Furthermore, CGD neutrophils are highly impaired in killing *Aspergillus* species and *Aspergillus* infections, especially with *A. nidulans*, consistently remain the most frequent cause of death in CGD. In addition, reconstitution of the NADPH oxidase by gene transfer restores NET formation and killing of *Aspergillus* species.

On the basis of all these findings, the authors suggest that the inability of CGD neutrophils to form NETs is the main reason for *Aspergillus* infections in patients with CGD [1]. However, the observation that completely myeloperoxidase (MPO)-deficient patients normally do not suffer from *Aspergillus* infections challenges this conclusion [2]. Completely MPO deficient neutrophils are also entirely unable to form NETs at least when stimulated with PMA or *Candida* species. [3]. Normally, MPO further metabolizes ROS produced by the NADPH oxidase to form different kinds of ROS such as hypochlorous acid. Whether such MPO metabolites or another ROS/MPO dependent mechanism is ultimately necessary for NET formation remains unknown. So far, data suggest that it is not the NADPH oxidase, but an MPO-dependent downstream event that triggers NET formation [3].

MPO deficiency is relatively frequent (1 in 2000-4000 in Europe and the United States), and complete MPO deficiency represents a considerable portion of these [2]. Therefore, complete MPO deficiency is much more frequent than CGD (~ 1 in 200,000 births). *Candida albicans* infections are somewhat more common in patients with MPO deficiency than in the normal population especially when additional influences such as diabetes contribute to susceptibility. However, according to the information available, susceptibility to *Aspergillus* species, in an order of magnitude comparable to CGD, can be excluded in patients who are completely deficient in MPO despite the inability of their neutrophils to form NETs.

It is very unlikely that completely MPO-deficient neutrophils form NETs only when stimulated with *Aspergillus* species, but not when stimulated with PMA or *Candida* species. In one study, MPO-deficient neutrophils even failed to damage *A fumigatus* hyphae in vitro [4]. However, mice with MPO deficiency showed only a slightly delayed clearance of *A. fumigatus* from the lungs [5]. In contrast, mice with CGD survived *Aspergillus* infections only for a short time [6]. Therefore, mammalian neutrophils and/or other phagocytes have most likely redundant abilities to kill *Aspergillus* species that are all ROS-dependent and therefore defective in CGD. However, details remain to be elucidated by appropriate experiments.

Disclosure of potential conflict of interest: the authors have declared that they have no conflict of interest.

References

1. Bianchi M, Niemiec MJ, Siler U, Urban CF, Reichenbach J. Restoration of anti-*Aspergillus* defense by neutrophil extracellular traps in human chronic granulomatous disease after gene therapy is calprotectin-dependent. J Allergy Clin Immunol 2011;127:1243-52.

2. Lekstrom-Himes JA, Gallin JI. Immunodeficiency diseases caused by defects in phagocytes. N Engl J Med 2000; 343(23):1703-14.

3. Metzler KD, Fuchs TA, Nauseef WM, Reumaux D, Roesler J, Schulze I et al. Myeloperoxidase is required for neutrophil extracellular trap formation: implications for innate immunity. Blood 2011; 117(3):953-9.

4. Diamond RD, Clark RA. Damage to Aspergillus fumigatus and Rhizopus oryzae hyphae by oxidative and nonoxidative microbicidal products of human neutrophils in vitro. Infect Immun 1982; 38(2):487-95.

5. Aratani Y, Kura F, Watanabe H, Akagawa H, Takano Y, Suzuki K et al. Differential host susceptibility to pulmonary infections with bacteria and fungi in mice deficient in myeloperoxidase. J Infect Dis 2000; 182(4):1276-9.

6. Aratani Y, Kura F, Watanabe H, Akagawa H, Takano Y, Suzuki K et al. Relative contributions of myeloperoxidase and NADPH-oxidase to the early host defense against pulmonary infections with Candida albicans and Aspergillus fumigatus. Med Mycol 2002;40(6):557-63.

Response sent to *JACI* regarding the correspondence: Roesler J, and Rösen-Wolff A. Redundant ability of phagocytes to kill *Aspergillus* species. *J Allergy Clin Immunol* 2011; online June 20

CORRESPONDENCE

Reply: *Aspergillus* infection in CGD and MPO deficiency

Matteo Bianchi, MSc,[a] Maria J. Niemiec, MSc,[b] Ulrich Siler, PhD,[a] Constantin F. Urban, PhD,[b‡] and Janine Reichenbach, MD,[a‡]

[a] Division of Immunology/Hematology/BMT, University Children's Hospital Zurich, Switzerland

[b] Antifungal Immunity Group, Molecular Biology Department, Laboratory for Molecular Infection Medicine Sweden, Umeå University, Umeå, Sweden

Corresponding Author:
Janine Reichenbach
Division of Immunology/Hematology/BMT
University Children's Hospital Zurich
Steinwiesstrasse 75, 8032 Zurich, Switzerland
Phone: +41 44 266 7311, Fax: +41 44 266 7914
e-mail: janine.reichenbach@kispi.uzh.ch

Published in: *Journal of Allergy and Clinical Immunology* 2011;128:687-88
Own contribution: M.B. contributed to the writing of the correspondence.

CORRESPONDENCE

Reply

To the Editor:

We thank Roesler and Rösen-Wolff for their comments[1] regarding our recent publication titled "Restoration of anti-Aspergillus defense by neutrophil extracellular traps in human chronic granulomatous disease after gene therapy is calprotectin-dependent" in the *Journal of Allergy and Clinical Immunology*.[2] They raise several interesting points, questioning our suggestion that formation of neutrophil extracellular traps (NETs) and NETs-mediated killing of Aspergillus species might contribute to clearance in healthy individuals. We concluded this, because neutrophils from patients with chronic granulomatous disease (CGD) do not release NETs on contact with *Aspergillus* and as a consequence killing is impaired. In the following text, we discuss the comments of Roesler and Rösen-Wolff.

Data on myeloperoxidase (MPO) deficiency (MIM #254600) were first published in 1966 by the name of Alius-Grignaschi anomaly, which was defined as total hereditary peroxidase deficiency of neutrophils and monocytes in a healthy person in this publication. Epidemiological data on this rare disease are scarce and controversial (reference list available on request) incidence rates in the United States and Europe range from 1 case in 2,000 (partial MPO deficiency) to 1 in 4,000 (complete MPO deficiency) population and from 1 in 17,500 (partial MPO deficiency) to 1 in 57,135 (complete MPO deficiency) population in Japan. Some studies do not differentiate between partial and complete deficiency, whereas others report results only from screening of a small collective originating from a restricted geographical area; in addition, a uniform definition of complete MPO deficiency is lacking (eg, null mutation with the absence of protein expression vs hypomorphic mutation with impaired enzymatic function). Information on clinical consequences of complete MPO deficiency, especially with regard to infection, is rare, and a review is lacking to date (reference list available on request). Indeed, a more definitive epidemiological survey would be an important contribution to understanding MPO deficiency.

Therefore, it is difficult to draw any conclusion as to susceptibility to infection of patients completely deficient in MPO. In addition, exposure to infectious agents might vary from one geographical region to another. As Roesler and Rösen-Wolff state correctly, *Candida* infections seem common in patients completely deficient in MPO[3]; however, disseminated infection with *Aspergillus flavus* and pulmonary infection with *Aspergillus fumigatus*[3,4] as well as invasive bacterial infection with *Staphylococcus aureus* and *Legionella* have been described (references available on request). Thus, the increased susceptibility to infection might be linked to the absence of NETs formation in individuals completely deficient in MPO and in patients with CGD,[2,3] although definitive in vivo evidence is obviously lacking.

Given the inability of completely MPO-deficient neutrophils to form NETs and some published evidence of susceptibility to infection, identification and a detailed clinical workup of further patients with complete MPO deficiency seems to be warranted in order to clarify its status as a primary immunodeficiency according to the WHO classification (Primary immunodeficiency diseases: an update from the International Union of Immunological Societies Primary Immunodeficiency Diseases Classification Committee Meeting in Budapest, 2005). Such a study would need to analyze genotype-phenotype correlation and susceptibility to infection and optimally comprise in vitro analyses of NETs formation on stimulation with identified disease-relevant pathogens.

Susceptibility in murine models should be interpreted carefully, as clinical phenotypes in mice and men do not always correlate. Interestingly, earlier studies with human phagocytes showed that MPO-deficient as well as CGD neutrophils damage *Aspergillus fumigatus* hyphae equally inefficiently.[5,6] Notably, hydrogen peroxide (a product of the nicotinamide adenine dinucleotide phosphate [NADPH] oxidase) or hypochlorous acid (a product of MPO) alone damaged *Aspergillus* hyphae only in concentrations ≥ 1 mM,[5] and extracellular products of MPO did not rescue NETs formation in response to *Candida albicans* stimulation in patients completely deficient in MPO,[3] suggesting that MPO acts cell-intrinsically, independent of its enzymatic activity, most likely by driving chromatin decondensation during NETs formation.[7]

To the best of our knowledge, NETs formation and NETs killing of *Aspergillus* species conidia or hyphae have not been analyzed yet in patients completely deficient in MPO. To address Roesler's and Rösen-Wolff's concerns, neutrophils of patients completely deficient in MPO should be analyzed parallely with those of patients with CGD and healthy controls and by adding products of MPO (eg, hypochlorous acid, histamine chloramines) and/or NADPH oxidase (eg, H_2O_2) to neutrophils.

We therefore conclude that according to current knowledge, both NADPH oxidase and MPO are required for fungus-induced NETs formation. However, the apparent divergence of major pathogens related to CGD and complete MPO deficiency, concluded from scarce, thus potentially unreliable, epidemiological data in complete MPO deficiency, remains to be defined.

Disclosure of potential conflict of interest: J. Reichenbach and M. Bianchi have received research support from the Chronic Granulomatous Disorders (CGD) Research Trust. The rest of the authors have declared that they have no conflict of interest.

References

1. Roesler J, Rösen-Wolff A. Redundant ability of phagocytes to kill aspergillus spp. J Allergy Clin Immunol 2011; In press (online June 20).
2. Bianchi M, Niemiec MJ, Siler U, Urban CF, Reichenbach J. Restoration of anti-Aspergillus defense by neutrophil extracellular traps in human chronic granulomatous disease after gene therapy is calprotectin-dependent. J Allergy Clin Immunol 2011; 127:1243-52.
3. Metzler KD, Fuchs TA, Nauseef WM, Reumaux D, Roesler J, Schulze I, et al. Myeloperoxidase is required for neutrophil extracellular trap formation: implications for innate immunity. Blood 2011; 117:953-9.
4. Chiang AK, Chan GC, Ma SK, Ng YK, Ha SY, Lau YL. Disseminated fungal infection associated with myeloperoxidase deficiency in a premature neonate. Pediatr Infect Dis J 2000; 19:1027-9.
5. Diamond RD, Clark RA. Damage to Aspergillus fumigatus and Rhizopus oryzae hyphae by oxidative and nonoxidative microbicidal products of human neutrophils in vitro. Infect Immun 1982; 38:487-95.
6. Rex JH, Bennett JE, Gallin JI, Malech HL, Melnick DA. Normal and deficient neutrophils can cooperate to damage Aspergillus fumigatus hyphae. J Infect Dis 1990; 162:523-8.
7. Papayannopoulos V, Metzler KD, Hakkim A, Zychlinsky A. Neutrophil elastase and myeloperoxidase regulate the formation of neutrophil extracellular traps. J Cell Biol 2010; 191:677-91.

CHAPTER 3

In vitro models for NADPH oxidase function to evaluate efficacy of new myelospecific γ-retroviral SIN vectors for X-CGD gene therapy

Matteo Bianchi, Ulrich Siler, Reinhard A. Seger, Janine Reichenbach

Division of Immunology/Hematology/BMT, University Children's Hospital Zurich, Switzerland

Corresponding Author:
Janine Reichenbach
Division of Immunology/Hematology/BMT
University Children's Hospital Zurich
Steinwiesstrasse 75, 8032 Zurich, Switzerland
Phone: +41 44 266 7311, Fax: +41 44 266 7914
e-mail: janine.reichenbach@kispi.uzh.ch

Written as manuscript for publication
Own contribution: M.B. performed the experiments, analyzed data, prepared all figures and wrote the manuscript.

Abstract

Chronic granulomatous disease (CGD) is a primary immunodeficiency, caused by impaired function of phagocyte nicotinamide adenine dinucleotide phosphate (NADPH) oxidase, resulting in poor antimicrobial activity of neutrophils, including the inability to generate neutrophil extracellular traps (NETs). Recent phase I/II clinical X-CGD gene therapy (GT) trials with a gamma (γ)-retroviral long terminal repeat (LTR)-driven gp91^{phox} vector proved efficacy of this treatment. Major drawbacks in these trials were transgene silencing and transactivation of oncogenes. Therefore new vectors with improved efficacy and safety are needed. We evaluated *in vitro* the efficacy of next generation γ-retroviral self inactivating (SIN) vectors, in which gp91^{phox} expression is driven by cloned internal human myelospecific promoters. In an X-CGD (gp91$^{phox-/-}$) murine model, NADPH oxidase reconstitution in transduced and differentiated human PLB-985 X-CGD and human lineage negative (Lin⁻) X-CGD hematopoietic stem cells (HSC) highlighted two γ-retroviral SIN vectors, with MRP8 and miRNA-223 promoters, as candidates for future human X-CGD GT. In this model we showed that NET formation cannot be restored in PLB-985 X-CGD cells, even with functional NADPH oxidase. Only gene corrected Lin⁻ X-CGD HSC could form NETs, suggesting that PLB-985 X-CGD lack signaling leading to NET cell death, likely due to immortalization of this cell line.

Introduction

Chronic granulomatous disease (CGD) is a primary immunodeficiency in which one of the genes encoding phagocyte nicotinamide adenine dinucleotide phosphate (NADPH) oxidase subunits is mutated, resulting in absent, or in rare cases limited, reactive oxygen species (ROS) production and poor microbial killing. Patients with CGD are susceptible to recurrent, severe, life-threatening bacterial and fungal infections [1]. CGD is a suitable disease for hematopoietic stem cell (HSC)-targeted gene therapy (GT), as it arises from single gene defects in HSCs, and as the five defective genes have been cloned.

As shown by recent human phase I/II clinical trials (in 2004 [2] and 2007 [3]), GT for X-linked gp91^{phox}-deficient CGD (X-CGD) with a monocistronic gamma (γ)-retroviral long terminal repeat (LTR)-driven vector (SF71gp91^{phox}), resulted in long-lasting engraftment of gene-corrected cells with therapeutic relevant levels of superoxide production, and lead to eradication of therapy refractory bacterial and fungal infections. However, unexpected later transgene silencing and oncogene transactivation leading to monosomy 7 and myelodysplastic syndrome (MDS) [2,4], mandate development of new vectors with sustained efficacy and safety.

Next generation self inactivating γ-retroviral (SIN) vectors have been developed (in collaboration with Dr. Manuel Grez, Georg Speyer Haus, Frankfurt) that contain a deletion in the U3 region of the 3' viral LTR, resulting in the inactivation of the viral promoter/enhancer elements after reverse transcription [5,6]. Transgene expression is driven by internal tissue-specific human promoters (SP107 [7], MRP8 [8], miRNA-223 [9]), which lack enhancer activity and should not be prone to silencing, and are cloned 5' to the gp91^{phox} open reading frame (ORF).

We studied all of these new vectors for *ex vivo* myelospecificity of gp91^{phox} expression and reconstitution of NADPH oxidase function by *in vitro* transduction and differentiation of human PLB-985 X-CGD cells and human lineage negative (Lin⁻) HSC from an X-CGD donor. For the first time we included functional analysis of NADPH oxidase-dependent neutrophil extracellular trap (NET) formation and NET-*Aspergillus* killing [3,10] in the evaluation of vector efficacy. This is a better readout than standard nitrobluetetrazolium (NBT) and dihydrorhodamine-123 (DHR) assays, which only allow evaluation of ROS production.

Here we show that, among the myelospecific vectors, MRP8 and miRNA-223 promoters are the most efficient in driving gp91^{phox} expression when used at multiplicity of infection (MOI) of 1, leading to sufficient ROS production for NET formation.

Materials and methods

Cell culture

If not stated otherwise, HEK 293T-based Phoenix A (ΦNX A) cells (Nolan Lab) were grown in DMEM (PAA) + 10% fetal bovine serum (FBS) (PAA) + 2% Penicillin/Streptomycin (PS) (Gibco). Myeloid PLB-985 X-CGD [11,12] cells were kept undifferentiated in RPMI (PAA) + 10% FBS + 2% PS, or differentiated to granulocyte-like cells with 0.5% dimethylformamide (DMF) (Sigma) + 2.5% FBS (1 week) [13]. Differentiation was confirmed by FACS by CD11b⁺ expression. Human Lin⁻ HSC from control and X-CGD donors were isolated from apheresis-mobilized peripheral HSC and bone marrow, respectively, by AutoMACS (Miltenyi) using Lineage Cell Depletion Kit (Miltenyi) and expanded undifferentiated in X-Vivo 10 (Lonza) + 2 mM L-glutamine (Gibco) + 2% embryonic stem screened FBS (Hyclone) + 1% PS + 150 ng/ml SCF + 150 ng/ml Flt3L + 50 ng/ml IL-3 + 50 ng/ml TPO, or differentiated to granulocyte-like cells over three weeks in RPMI + 2 mM L-glutamine + 10% FBS + 1% PS + 50 ng/ml SCF (only first week) + 10 ng/ml Flt3L (only first two weeks) + 20 ng/ml G-CSF (during all three weeks). Differentiation was confirmed by FACS by CD15⁺ and CD11b⁺ expression.

Cloning of γ-retroviral SIN vectors and production of viral particles

The γ-retroviral pSER11M8delN91 SIN vector (kindly provided by Dr. Manuel Grez, Georg Speyer Haus, Frankfurt) was used as template for all constructs. This vector contains the internal human MRP8 promoter 5' and the Woodchuck hepatitis virus post-transcriptional element (WPRE) 3' of the gp91^{phox} transgene, and a deleted 3' U3 LTR region. The constitutive active SFFV promoter and GFP sequences were amplified from the "pSER11.SF.GFP.W" vector (kindly provided by Dr. M. Grez). The myelospecific SP107 (SP for synthetic promoter) and miRNA-223 promoters were derived from "Lenti-sp107-GFP" [7] (kindly provided by Senlin Li, University of Texas, Department of Medicine, USA) and "223-1gpsW" vectors (kindly provided by Dr. M. Grez). Recombinant plasmids were introduced by heat shock into *E. coli* Top-10 competent cells (Invitrogen). DNA extraction and purification were performed according to standard procedures using commercial kits (Quiagen, Macherey-Nagel). All cloned vectors were checked by restriction digestion and agarose gel electrophoresis. Sequences were confirmed by DNA sequencing (ABI PRISM 3130xl Genetic Analyzer, Applied Biosystems).

Amphotropic infectious particles were generated by transient transfection of ΦNX A cells. In 15 cm plates, 70-80% confluent ΦNX A cell layers (in 15 ml media) were co-transfected with 20 μg pUMVC (γ-retroviral Gag-Pol), 10 μg VSV-G (Env) and 80 μg transgene vectors by 4 μl TransIT-293 reagent (Mirus) per μg of total DNA. Viral particle containing supernatants were collected 24 and 48 h after transfection, sterile filtered through 0.45 μm PVDF membranes (Millipore) and concentrated from 15 ml to 500 μl using 100 kDa cutoff centrifugal filter units (Amicon-Ultra 15, Millipore). Concentrated viral supernatant was aliquoted (10 μl) and kept at -80°C until use.

Transduction of PLB-985 X-CGD and human Lin⁻ HSC X-CGD cells

For titration, PLB-985 X-CGD cells were transduced by spinoculation (90 min, 2500 RPM, 32°C) with serial dilutions of viral supernatants in presence of 8μg/ml protamine sulfate (Sigma). Transduction rate was measured 3 days later by flow cytometry (FACS Calibur, BD Biosciences). As transduction efficiencies above 20 % result in a significant portion of cells with more than one integration [14] we calculated the titers from those dilutions resulting in a maximum of 20% transgene positive cells, using the formula: Titer (Transducing Units (TU)/ml) = seeded cell number x dilution factor of viral supernatant x (transgene positive cells (%)/100%). For transgene expression analysis, transduced PLB-985 X-CGD cells were either kept undifferentiated or differentiated to granulocyte-like cells in 0.5% DMF. For functional analysis, MOI 1 transduced cells (< 20% transduction rate) were first stained with 10 μg/ml $gp91^{phox}$-FITC and sorted with anti-FITC beads by AutoMACS (Miltenyi) to obtain $gp91^{phox+}$ cells, then either expanded undifferentiated or differentiated to granulocyte-like cells [15].

Human Lin⁻ HSC were transduced on 10 μg/cm² Retronectin (TaKaRa) precoated plates. Coating was performed following product instructions. Afterwards, γ-retroviral supernatants were first loaded onto Retronectin-coated plates by centrifugation (3000 RPM for 30 min at 4°C), then Lin⁻ HSC were seeded, centrifuged 5 min at 300 g and expanded undifferentiated for up to 7 days or differentiated to granulocyte-like cells over 3 weeks for further functional analysis.

NADPH oxidase function analysis and NET formation

NADPH oxidase activity was determined by DHR oxidation [16] and by nitrobluetetrazolium (NBT) reduction [17] as described. The amount of O_2- produced by phorbol 12-myristate 13-acetate (PMA) stimulated cells was quantified by Fe^{3+} cytochrome c reduction as described [2,18]. NET formation was analyzed by 6 h incubation of $5x10^4$ cells/well (standard cell culture 96 well plate, TPP) in RPMI (without phenol red) + 5 mM Hepes + 40 nM PMA. NETs were fixed in 4% paraformaldehyde, stained with 5 µM Sytox green (Invitrogen) and visualized using an inverted light microscope (Leica DM IL HC Fluo) with a 100 W mercury lamp (106z, Leica) and a GFP filter (GFP Filter Cube, Leica). Images were taken using a CCD camera (DFC 480 R2, Leica) coupled to the microscope.

Results

Next generation γ-retroviral SIN vectors

pSF71gp91, the LTR driven vector used in previous GT clinical trials for treatment of X-CGD [2,3], will be replaced by new generation γ-retroviral SIN vectors (Figure 1) in future GT clinical trials. New vectors have a codon optimized $gp91^{phox}$ sequence, deleted 3' LTR enhancer, and TATA box and internal myelospecific promoters to drive transgene expression. All vectors were cloned using pSER11M8delN91 as template. Two myelospecific promoters (SP107 and miRNA-223) and for comparison a strong constitutively active viral promoter (SFFV) were cloned. $gp91^{phox}$ was replaced by GFP as control vector and for setup experiments. Viral titers were initially low in PLB-985 X-CGD cells, therefore intense optimization work using GFP vector was performed resulting in a protocol which in summary required: adjusted plasmid amounts, lipid-mediated co-transfection and viral particles concentration. A 1000-fold increase of titers, from 10^4 TU/ml to up to 10^7 TU/ml, was achieved following this protocol, with differences between vectors (Table 1).

Table 1. Viral titers

Vector	Titers (TU/ml)
pSF71gp91	$3.40x10^6$
pSER11SFFVGFP	$1.20x10^7$
pSER11SFFVgp91	$1.15x10^7$
pSER11SP107delN91	$4.08x10^6$
pSER11M8delN91	$1.59x10^7$
pSER11miR223delN91	$1.02x10^7$

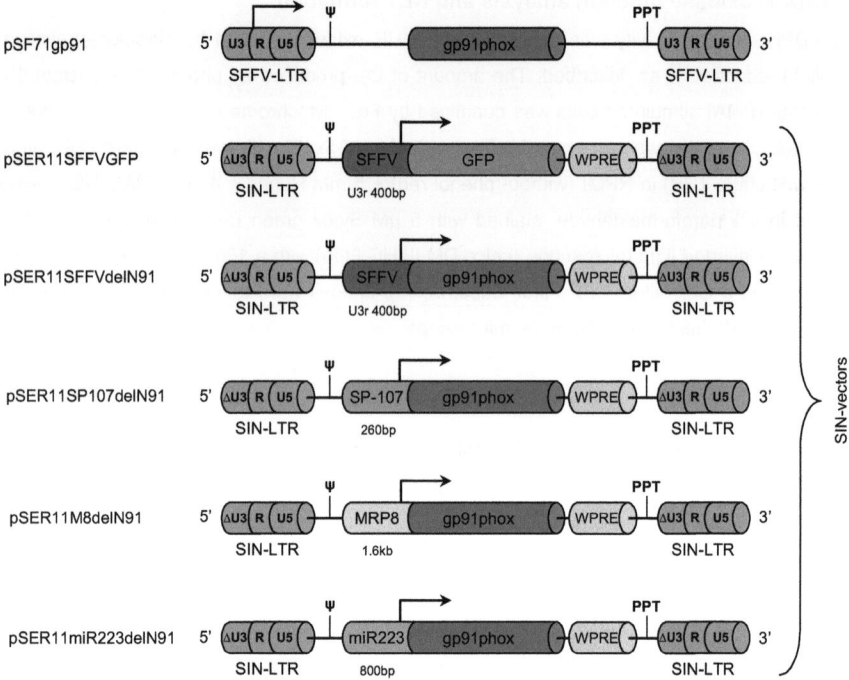

Figure 1. Next generation γ-retroviral SIN vectors. Previously used LTR-driven vector (pSF71gp91) and new generation of cloned γ-retroviral SIN vectors are shown. SIN vectors contain the Woodchuck hepatitis virus post-transcriptional element (WPRE) 3' of the transgene and a deleted enhancer and TATA box in the 3' LTR U3 region. Transgene expression is driven by internal promoters. SFFV = Spleen Focus Forming Virus promoter; SP107 = synthetic promoter composed by nine PU.1 transcription factors binding sites; MRP8 = S100A8 (Calgranulin A) promoter; miR223 = micro RNA 223 promoter.

Myelospecificity of next generation γ-retroviral SIN vectors

To screen for myelospecificity, premyeloid PLB-985 X-CGD cells were used. Cells were transduced with all gp91^{phox} encoding γ-retroviral vector constructs, then differentiated to granulocyte-like cells. CD11b staining showed good differentiation after 1 week in 0.5% DMF, more than 95% of the cells were CD11b$^+$ as analyzed by FACS. There was no difference in gp91^{phox} expression between undifferentiated and differentiated PLB-985 X-CGD cells, when transgene expression was driven by the strong and non-myelospecific SFFV promoter (SF71gp91 and SFFVgp91, Figure 2). Unexpectedly, gp91^{phox} expression was 50% lower in differentiated cells with SP107gp91 vector, with very low "mean fluorescence intensity" (MFI) of the gp91^{phox+} cell population, indicating that only low transgene expression could be obtained by using the SP107 promoter. MRP8gp91 and miR223gp91 vectors showed more promising results with an increase of around 52% and

14% of gp91^{phox} expressing neutrophils-like cells and 6% and 18% increase of transgene expression (MOI), respectively, after cell differentiation (Table 2).

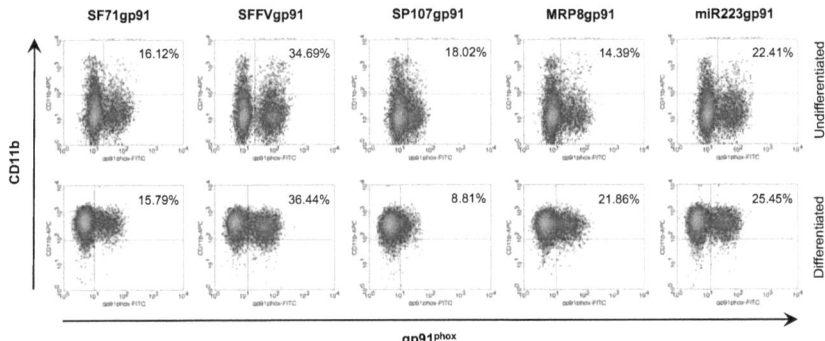

Figure 2. Myelospecificity of γ-retroviral SIN vectors. Myeloid PLB-985 X-CGD cells were transduced with all gp91^{phox} γ-retroviral vectors and then differentiated to granulocyte-like cells in 0.5% DMF. gp91^{phox} expression was compared in undifferentiated (CD11b⁻) and differentiated (CD11b⁺) cells. SF71gp91 and SFFVgp91 are used as constitutively active promoter/vector controls. Among other promoters, MRP8 and miR223 showed the highest myelospecific gp91^{phox} expression after differentiation, with an increase of around 52% and 14% of transgene expressing cells, and 6% and 18% increase of transgene expression (MOI), respectively. SP107 gp91^{phox} driven expression was the lowest.

As all promoters were already active in undifferentiated PLB-985 X-CDG cells, it is difficult to evaluate precisely any change in transgene expression in this cell line. As PLB-985 X-CDG is a premyeloid cell line, such cells are not enough undifferentiated, and the measured activity of the three myelospecific promoters in undifferentiated, yet premyeloid cells, is therefore not surprising.

Table 2. Myelospecificity of γ-retroviral vectors in PLB-985 X-CGD cells (¢)

Vector	% gp91^{phox+} in undif. ¢	% gp91^{phox+} in dif. ¢	Δ% ¢ upon dif.	Δ% MFI upon dif.
pSF71gp91	16.12	15.79	-2.05	-0.97
pSER11SFFVGFP	34.69	36.44	+5.04	-6.93
pSER11SP107delN91	18.02	8.81	-51.11	+6.58
pSER11M8delN91	14.39	21.86	+51.91	+5.97
pSER11miR223delN91	22.41	25.45	+13.57	+18.30

Abbreviations: undif = undifferentiated; dif = differentiated; ¢ = cells

Reconstitution of NADPH oxidase function in PLB-985 X-CGD cells

To better study functionality of gp91^{phox} γ-retroviral SIN vectors, MOI 1 transduced PLB-985 X-CGD cells were sorted after gp91^{phox} staining by AutoMACS, then expanded. One week culture in 0.5% DMF resulted in more than 90% differentiated cells, as shown by CD11b staining. After sorting, more than 80% of cells were CD11b/gp91$^{phox+/+}$ (Figure 3A).

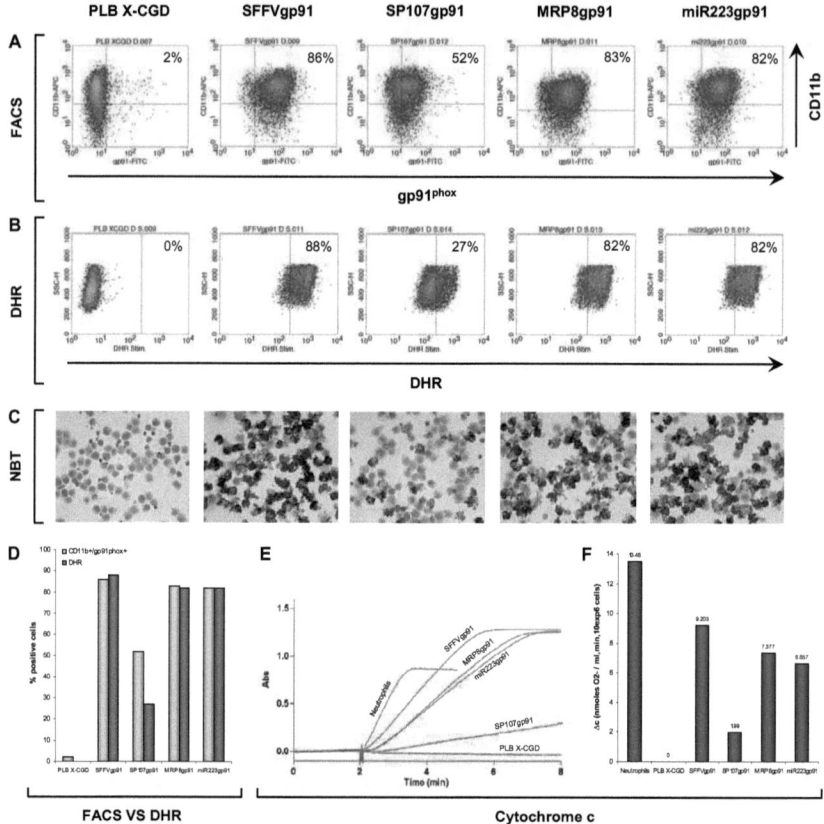

Figure 3. Reconstitution of NADPH oxidase function in PLB-985 X-CGD cells. MOI 1 transduced PLB-985 X-CGD gp91^{phox} expressing cells were sorted by AutoMACS and differentiated in 0.5% DMF (A). NADPH oxidase function was assessed by DHR (B) and NBT (C) assays, showing that MRP8gp91 and miR223gp91 were the most efficient vectors to reconstitute NADPH oxidase activity (the strong SFFVgp91 vector was used as control). SP107gp91 vector showed poor functionality: at MOI 1 not all gp91^{phox} expressing cells were able to reconstitute function (D). O_2^- quantification by cytochrome c Fe^{3+} reduction (E-F) confirmed DHR and NBT results: O_2^- production by MRP8gp91 and miR223gp91 were comparable to SFFVgp91 transduced cells, whereas in SP107gp91 transduced cells O_2^- production was insufficient.

Only cells transduced with the SP107gp91 vector resulted in lower CD11b/gp91$^{phox+/+}$ numbers (52%), as the sorted gp91^{phox+} population was less pure due to the very low gp91^{phox} expression driven by the SP107 promoter (Figure 2).

For testing functionality, ROS-production of transduced cells was first analyzed by DHR test (Figure 3B). Reconstitution of NADPH function was very efficient when SFFVgp91, MRP8gp91 and miR223gp91 vectors were used, with more than 80% DHR positive cells. The SP107gp91 vector could only partially restore ROS production, with only 27% DHR positive cells. These results were then confirmed by NBT assay, with a blue precipitate (formazan) being extensively formed with all vectors but SP107gp91 (Figure 3C). Comparison of CD11b/gp91$^{phox+/+}$ expression, DHR and NBT data showed good correlation between gp91^{phox} expression and NADPH oxidase function when cells were transduced with SFFVgp91, MRP8gp91 and miR223gp91 vectors (Figure 3D), i.e. expression of a single transgene copy/cell driven by these promoters is enough to establish sufficient functionality. This was not the case for the SP107gp91 vector.

ROS production was then quantified by cytochrome c Fe^{3+} reduction (Figure 3E), showing that O_2^- production was, as expected, more than 3.5-fold lower in SP107gp91 transduced cells compared to cells transduced with all other vectors (Figure 3F, values normalized to CD11b/gp91$^{phox+/+}$ expression). MRP8gp91 and miR223gp91 vectors behaved again very similar to each other, with only slightly lower O_2^- production compared to the strong SFFVgp91 vector. Normal healthy neutrophils were used as reference.

NET formation in transduced PLB-985 X-CGD cells

Activated neutrophils kill microbes via phagocytosis and by extracellular mechanisms including Neutrophil Extracellular Traps (NETs). NETs are composed of chromatin (DNA and histones) decorated with granular proteins [19]. These structures bind bacteria [19] and fungi [20] and provide a high local concentration of antimicrobial molecules. Generation of NETs requires the production of ROS by the NADPH oxidase, therefore is deficient in CGD [21].

To further test functional efficacy of next generation vectors, NET formation was therefore studied in all transduced PLB-985 X-CGD cells. After 6h stimulation with PMA, NET formation was visualized in all transduced and sorted cells by DNA stain and fluorescence microscopy analysis. No NETs were formed by transduced, NADPH oxidase expressing and ROS producing PLB-985 X-CGD cells, likely because of the immortalization characteristics of this cell line. Figure 4 shows normal healthy neutrophils as control, with compact and multilobated nuclei while unstimulated, and decondensed NET-DNA upon PMA stimulation. In contrast, PLB-985 X-CGD cells transduced with the SFFVgp91 vector (as representative result only the strongest vector is shown) did not show any significant difference between unstimulated and stimulated cells. Only very few (< 5%) PMA activated cells showed a NET-like structure, and not more than 80% as it was expected from DHR and NBT data (Figure 3).

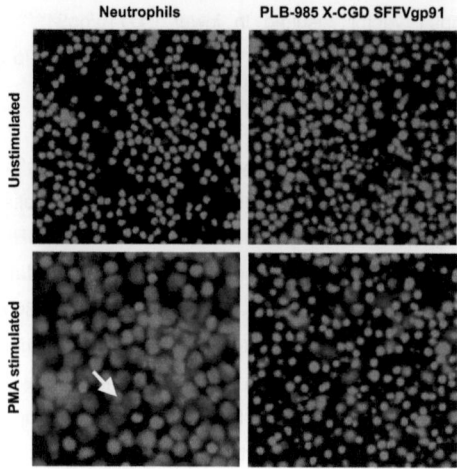

Figure 4. NET induction in transduced PLB-985 X-CGD cells. MOI 1 transduced PLB-985 X-CGD gp91^{phox} expressing cells were sorted by AutoMACS and differentiated in 0.5% DMF. NETs were induced by 6 h incubation with 40 nM PMA, NET-DNA was stained with 5 µM Sytox green after fixation in 4% paraformaldehyde and visualized by fluorescence microscopy. Healthy neutrophils were used as control, showing full NET formation after PMA stimulation (left lower panel, arrow shows a representative NET-cell). In contrast, PLB-985 X-CGD cells transduced with the strong SFFVgp91 vector were not able to make NETs efficiently (< 5% right lower panel).

ROS production and reconstitution of NET formation in Lin⁻ X-CGD HSC

To overcome above mentioned problem with immortalized PLB-985 X-CGD cells, the same functional NET assay was performed using human Lin⁻ HSC isolated from apheresis-mobilized peripheral HSC of a healthy donor and Lin⁻ HSC isolated from bone marrow of an X-CGD patient. Lin⁻ HSC were pre-stimulated to induce cell division (as γ-retroviral vectors do not integrate in resting cells), transduced with MOI 1 on Retronectin and further cultured (Figure 5).

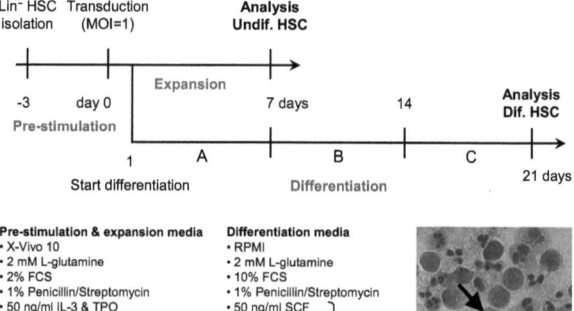

Figure 5. Culture and differentiation of human Lin⁻ HSC. Human Lin⁻ HSC were isolated from apheresis-mobilized peripheral HSC of an healthy donor and bone marrow of an X-CGD patient by AutoMACS. After 3 days pre-stimulation, cells were transduced (at MOI 1) and expanded undifferentiated for 1 week or differentiated to granulocyte-like cells over 3 weeks. Differentiation was determined by CD15/CD11b$^{+/+}$ expression and by microscopical analysis of multilobated nuclei (arrow in lower right panel).

Due to lower cell numbers available and limited *in vitro* expansion possibilities (compared to PLB-985 X-CGD cells, Lin⁻ HSC are slow dividing cells and cannot be expanded indefinitely as they easily start to self-differentiate in different cellular subtypes), transduced X-CGD Lin⁻

HSC were not sorted for gp91^{phox} expression. Differentiation was assessed by FACS analysis of CD15$^+$ and CD11b$^+$ expression, and by microscopical analysis for presence of multilobated nuclei. After 3 weeks differentiation, neutrophil-like cell (CD15/CD11b$^{+/+}$) formation rate was estimated to be around 45% by FACS (not shown), confirmed by microscopical evaluation of nuclear shape (Figure 5). As expected, FACS analysis of gp91^{phox} expression in transduced cells showed that transduction of X-CGD Lin$^-$ HSC cells with γ-retroviral vectors was not as efficient as in PLB-985 X-CGD cells, as cell division of Lin$^-$ HSC is not as fast as cell division in cell lines. Transduction rate was around 11% with SFFVgp91 vector, 4-6% with MRP8gp91 and miR223gp91 vectors and only 1% with SP107gp91 vector (not shown), confirming the low transduction efficiency of the SP107 promoter. NADPH oxidase dependent ROS production was restored after transduction and differentiation of X-CGD Lin$^-$ HSC, as shown by DHR and NBT tests (Figure 6A-B), in good correlation with CD15/CD11b$^{+/+}$ and gp91^{phox+} expression. In contrast to transduced PLB-985 X-CGD cells, PMA stimulation and DNA staining evidenced NET formation in control, SFFVgp91, MRP8gp91 and miR223gp91 transduced and differentiated X-CGD Lin$^-$ HSC cells, as shown by fluorescence microscopy (Figure 6C). Extent of NET formation was comparable to NBT results (Figure 6B).

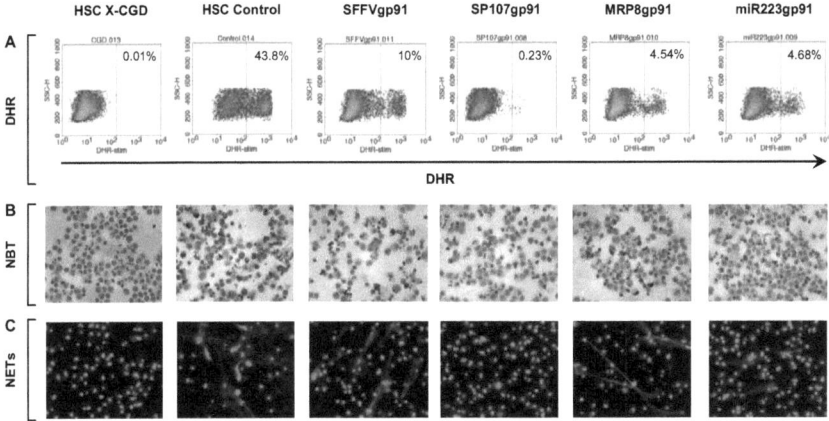

Figure 6. Reconstitution of NADPH oxidase dependent ROS production and NET formation in human Lin$^-$ HSC. Untransduced normal healthy human Lin$^-$ HSC, and untransduced and MOI 1 transduced human X-CGD Lin$^-$ HSC were differentiated to granulocyte-like cells over 3 weeks. NADPH oxidase dependent ROS production was assessed by DHR (A) and NBT (B) assays, showing that MRP8gp91 and miR223gp91 were most efficient (the strong SFFVgp91 vector was used as reference, together with untransduced healthy Lin$^-$ HSC). SP107gp91 vector was unable to restore sufficient ROS production. NETs were induced by 6 h incubation with 40 nM PMA, NET-DNA was stained with 5 μM Sytox green after fixation in 4% paraformaldehyde and visualized by fluorescence microscopy (C). In contrast to transduced PLB-985 X-CGD gp91^{phox+} cells, NETs were formed by transduced and differentiated Lin$^-$ HSC, in good correlation with DHR and NBT results.

Discussion

CGD is an attractive target for GT because it is caused by a single gene defect, it can be reproduced *in vitro* and because the genetic correction does not need to be complete to provide full protection against infections [22]. Data from variant forms of CGD and from healthy female carriers of X-CGD with 10% normal neutrophils suggest that correction of the phenotype in a fraction of CGD cells could be sufficient to alleviate the symptoms of the disease [23].

γ-retroviral vectors have been largely studied over the years and phase I/II clinical trials of gene therapy against X-CGD proved the feasibility of this approach [2,3]. Side effects like silencing and transactivation [2,4] imposed the improvement of vector efficacy and safety, leading to the development of SIN vectors [5,6]. In contrast to lentiviral vectors, γ-retroviral vectors can only infect dividing cells. Titers of produced viral particles are lower using γ-retroviral vectors compared to lentiviral vectors, which is a problem transducing slowly dividing HSC. The first important step in our vector screen was therefore the improvement of viral titers with a 3-log factor, from 10^4 TU/ml to up to 10^7 TU/ml, by optimization of plasmid concentrations utilized for virus production, lipid transfection and concentration of viral particles. Clinically, high viral titers are important because they allow shorter *ex vivo* culture as they need less HSC transduction rounds, which may help preserve the engraftment capacity of the HSC for long term correction [24].

Among the three myelospecific promoter containing vectors tested with regard to transgene expression and NADPH oxidase function, vectors containing MRP8 and miRNA-223 promoters are the most promising for future clinical use. Both vectors restored strong $gp91^{phox}$ expression and O_2^- production in transduced and differentiated PLB-985 X-CGD cells, comparable to a strong SFFV-driven SIN vector. The fact that NADPH oxidase function was fully restored at MOI of 1 is of importance with regards to vector safety, as multiple integrations have to be avoided to lower the risk of transactivational adverse events. SP107gp91 vector showed the lowest NADPH oxidase reconstitution potential in the studied models. $gp91^{phox}$ expression driven by the SP107 promoter was too low for sufficient ROS production when used at MOI of 1. O_2^- production was 3.5-fold lower compared to the other myelospecific vectors, barely enough to give a positive DHR or NBT signal in the few corrected cells.

Myelospecificity was difficult to assess when comparing undifferentiated to differentiated transduced PLB-985 X-CGD cells as PLB-985 is a pre-myeloid cell line [11], already partially differentiated into the monocytic/neutrophilic direction. It is therefore not surprising that all myelospecific promoters used in our screen were already active in undifferentiated transduced PLB-985 X-CGD cells. After differentiation, only MRP8 and miRNA-223 promoters showed slight improvements in gp91^{phox} expression, but the PLB-985 X-CGD model proved to be non-optimal to study promoter myelospecificity.

As we have previously shown the importance for NADPH oxidase-dependent NET formation in anti-*Aspergillus* defense [3,10] we included NET formation assays in our vector screen, allowing to study for the first time NADPH oxidase *functional reconstitution*. Unluckily transduced and differentiated PLB-985 X-CGD cells (and normal PLB-985 cells with functional NADPH oxidase) were not able to form NETs upon stimulation. This is presumably due to the immortalized characteristics of the cell line, precluding a cell death program as NETosis. We therefore performed the vector screen using human Lin⁻ HSC isolated from an X-CGD patient. DHR and NBT results confirmed efficient NADPH oxidase reconstitution in terms of ROS production with MRP8gp91 and miR223gp91 vectors. SP107gp91 was inefficient in correcting the defect. NET formation was efficiently restored after transduction. As mentioned above, the limiting factor using γ-retroviral vectors is their inability to infect non-dividing cells, as shown by the low percentage (4-10%) of transduced cells when using a single transduction round and MOI 1 protocol. For clinical purpose, at least a second transduction round would therefore be necessary in order to have > 10% NADPH oxidase functional neutrophils.

Myelospecificity was difficult to assess in transduced Lin⁻ HSC, as undifferentiated Lin⁻ HSC cannot be expanded efficiently and as stem cells properties are quickly lost during *in vitro* culture. Transgene expression in undifferentiated Lin⁻ HSC, even with myelospecific promoters, may be caused by the presence of unintegrated circular DNA [25] which could not get lost by dilution during cell divisions. Therefore there is a risk to compare multiple copies of unintegrated circular DNA in Lin⁻ HSC or progenitor cells with a single copy of integrated transgene in differentiated granulocyte-like cells (where unintegrated DNA get diluted by sustained cell division during the three weeks differentiation process).

In conclusion, our screens show that the SP107 promoter cannot be considered a candidate for human X-CGD GT, due to insufficient transgene expression. The use of this synthetic promoter, with nine repetitive PU.1 sites [7], would anyway need to be analyzed very carefully, as repetitive sequences within retroviral genomes are unstable and frequently deleted upon reverse transcription [26,27]. In contrast, vectors harboring the MRP8 and miRNA-223 promoters are eligible for further safety-centered studies, to be finally considered for future human X-CGD GT trials.

Authorship

Contribution: M.B. performed the experiments and analyzed the results together with U.S and J.R. The study was designed by J.R. and R.A.S.

Conflict-of-interest disclosure: The authors declare no competing financial interests.

References

1. Seger RA. Modern management of chronic granulomatous disease. Br J Haematol. 2008;140:255-266.
2. Ott MG, Schmidt M, Schwarzwaelder K, et al. Correction of X-linked chronic granulomatous disease by gene therapy, augmented by insertional activation of MDS1-EVI1, PRDM16 or SETBP1. Nat Med. 2006;12:401-409.
3. Bianchi M, Hakkim A, Brinkmann V, et al. Restoration of NET formation by gene therapy in CGD controls aspergillosis. Blood. 2009;114:2619-2622.
4. Stein S, Ott MG, Schultze-Strasser S, et al. Genomic instability and myelodysplasia with monosomy 7 consequent to EVI1 activation after gene therapy for chronic granulomatous disease. Nat Med. 2010;16:198-204.
5. Kraunus J, Schaumann DH, Meyer J, et al. Self-inactivating retroviral vectors with improved RNA processing. Gene Ther. 2004;11:1568-1578.
6. Schambach A, Mueller D, Galla M, et al. Overcoming promoter competition in packaging cells improves production of self-inactivating retroviral vectors. Gene Ther. 2006;13:1524-1533.
7. He W, Qiang M, Ma W, et al. Development of a synthetic promoter for macrophage gene therapy. Hum Gene Ther. 2006;17:949-959.
8. Kuwayama A, Kuruto R, Horie N, Takeishi K, Nozawa R. Appearance of nuclear factors that interact with genes for myeloid calcium binding proteins (MRP-8 and MRP-14) in differentiated HL-60 cells. Blood. 1993;81:3116-3121.
9. Johnnidis JB, Harris MH, Wheeler RT, et al. Regulation of progenitor cell proliferation and granulocyte function by microRNA-223. Nature. 2008;451:1125-1129.
10. Bianchi M, Niemiec MJ, Siler U, Urban CF, Reichenbach J. Restoration of anti-Aspergillus defense by neutrophil extracellular traps in human chronic granulomatous disease after gene therapy is calprotectin-dependent. J Allergy Clin Immunol. 2011;127:1243-1252 e1247.
11. Tucker KA, Lilly MB, Heck L, Jr., Rado TA. Characterization of a new human diploid myeloid leukemia cell line (PLB-985) with granulocytic and monocytic differentiating capacity. Blood. 1987;70:372-378.
12. Zhen L, King AA, Xiao Y, Chanock SJ, Orkin SH, Dinauer MC. Gene targeting of X chromosome-linked chronic granulomatous disease locus in a human myeloid leukemia cell line and rescue by expression of recombinant gp91phox. Proc Natl Acad Sci U S A. 1993;90:9832-9836.
13. Parkos CA, Dinauer MC, Walker LE, Allen RA, Jesaitis AJ, Orkin SH. Primary structure and unique expression of the 22-kilodalton light chain of human neutrophil cytochrome b. Proc Natl Acad Sci U S A. 1988;85:3319-3323.

14. Fehse B, Kustikova OS, Bubenheim M, Baum C. Pois(s)on--it's a question of dose. Gene Ther. 2004;11:879-881.
15. Ding C, Kume A, Bjorgvinsdottir H, Hawley RG, Pech N, Dinauer MC. High-level reconstitution of respiratory burst activity in a human X-linked chronic granulomatous disease (X-CGD) cell line and correction of murine X-CGD bone marrow cells by retroviral-mediated gene transfer of human gp91phox. Blood. 1996;88:1834-1840.
16. Vowells SJ, Sekhsaria S, Malech HL, Shalit M, Fleisher TA. Flow cytometric analysis of the granulocyte respiratory burst: a comparison study of fluorescent probes. J Immunol Methods. 1995;178:89-97.
17. Bohler MC, Seger RA, Mouy R, Vilmer E, Fischer A, Griscelli C. A study of 25 patients with chronic granulomatous disease: a new classification by correlating respiratory burst, cytochrome b, and flavoprotein. J Clin Immunol. 1986;6:136-145.
18. Mayo LA, Curnutte JT. Kinetic microplate assay for superoxide production by neutrophils and other phagocytic cells. Methods Enzymol. 1990;186:567-575.
19. Brinkmann V, Reichard U, Goosmann C, et al. Neutrophil extracellular traps kill bacteria. Science. 2004;303:1532-1535.
20. Urban CF, Reichard U, Brinkmann V, Zychlinsky A. Neutrophil extracellular traps capture and kill Candida albicans yeast and hyphal forms. Cell Microbiol. 2006;8:668-676.
21. Fuchs TA, Abed U, Goosmann C, et al. Novel cell death program leads to neutrophil extracellular traps. J Cell Biol. 2007;176:231-241.
22. Kang EM, Malech HL. Advances in treatment for chronic granulomatous disease. Immunol Res. 2009;43:77-84.
23. Barese CN, Goebel WS, Dinauer MC. Gene therapy for chronic granulomatous disease. Expert Opin Biol Ther. 2004;4:1423-1434.
24. Kennedy DR, McLellan K, Moore PF, Henthorn PS, Felsburg PJ. Effect of ex vivo culture of CD34+ bone marrow cells on immune reconstitution of XSCID dogs following allogeneic bone marrow transplantation. Biol Blood Marrow Transplant. 2009;15:662-670.
25. Haas DL, Case SS, Crooks GM, Kohn DB. Critical factors influencing stable transduction of human CD34(+) cells with HIV-1-derived lentiviral vectors. Mol Ther. 2000;2:71-80.
26. Junker U, Bohnlein E, Veres G. Genetic instability of a MoMLV-based antisense double-copy retroviral vector designed for HIV-1 gene therapy. Gene Ther. 1995;2:639-646.

27. Delviks KA, Pathak VK. Effect of distance between homologous sequences and 3' homology on the frequency of retroviral reverse transcriptase template switching. J Virol. 1999;73:7923-7932.

DISCUSSION AND OUTLOOK

Discussion and Outlook

CGD is a rare primary immunodeficiency, with an estimated frequency of 1 in 200'000 live births. Defective phagocyte NADPH oxidase function accounts for impaired ROS production and poor microbial killing, leading to recurrent life-threatening bacterial and fungal infections in patients with CGD. Overall mortality is a serious issue (among all patients who do not get BMT), with about 13-20% of the patients not surviving the disease despite adequate antibiotic and antifungal prophylaxis [1-3]. *Aspergillus spp* infections, which cause pneumonia and disseminated disease, are frequently refractory to antifungal therapy, treatment with interferon-γ, or granulocyte transfusions [4], and are the leading cause of death in CGD patients [3-6].

Objectives 1 and 2

Up to now, it was unclear how normal human neutrophils control the significantly larger *Aspergillus* hyphae which are too big for phagocytosis [7-13]. We elucidated the neutrophil immune response and mechanism against *Aspergillus* infection (*chapters 1 and 2*):

Objective 1: We demonstrated that the recently discovered NADPH oxidase dependent extracellular microbicidal pathway through NETs is acting efficiently against *A. nidulans* conidia and hyphae *in vitro*, and that restoration of NET formation by complementation of NADPH oxidase function by GT in a patient with X-linked CGD aided clearing severe invasive *A. nidulans* infection *in vivo (Bianchi et al, Blood 2009* [14]*)*.

Objective 2: The antifungal agent responsible for control of *A. nidulans* within NETs was identified to be the Zn^{2+}-chelator calprotectin. The calprotectin heteroduplex formed by subunits S100A8 and S100A9 belongs to the calcium-binding protein family of S100 proteins, and exerts strong microbistatic activity against a variety of microorganims *in vitro*. We established that calprotectin released in NETs displays concentration-dependent reversible and irreversible growth inhibitory activity against virulent *A. nidulans*: it was fungistatic at low concentrations (< 16ng/ml) and fungicidal at high concentrations (> 16ng/ml), which closer mimic the physiological situation in human abscesses (1-20 mg/ml found in such abscesses [15]). NETs therefore seem to be involved in disarming *A. nidulans* and may prevent further spreading by providing high local concentrations of calprotectin *(Bianchi et al, JACI 2011* [16]*)*.

In conclusion of objectives 1 and 2

As the presence of a functional NADPH oxidase in human neutrophils was shown to be an absolute requirement for *A. nidulans* growth inhibition and fungal starvation by NET-associated calprotectin, we propose that NET formation is an NADPH oxidase-dependent mechanism by which neutrophils secrete calprotectin into the extracellular compartment upon infection with *Aspergillus* hyphae. NET-associated calprotectin is concentrated with trapped *Aspergillus* for antifungal function. Unbound calprotectin may additionally serve for distinct functions, such as chemotaxis or inflammation.

Outlook concerning objectives 1 and 2

Since, for ethical reasons, the role of NETs in antimicrobial defense is difficult to monitor in humans *in vivo*, we are currently developing an *in vivo* infection model using CGD mice, involving subcutaneous injection of bioluminescent *Aspergillus* hyphae. NET formation and clinical as well as histological granuloma resolution will be monitored before and after GT in these mice. The aim of this assay is to correlate NET formation to NADPH oxidase gene expression and reconstitution of NADPH oxidase function to enable choice of the adequate next generation GT vector for future human clinical CGD GT trials.

GT approaches to cure CGD have been developed in the past years as an alternative to CGD patients needing HSCT, but lacking an HLA-matched HSC donor [5,14,16-20]. Despite clinical success with resolution of life-threatening therapy refractory infections in 4 patients (2 adults treated in Frankfurt University, and 2 children treated in our Division of Immunology/Hematology/BMT in Zurich) with X-linked CGD, two major drawbacks occurred during the clinical GT trial: silencing of the transgene with significant decline of NADPH oxidase function, and transactivation of oncogenes leading to clonal expansion and MDS in 3/4 patients [21].

Objective 3

Transactivation and silencing events impose the development of alternative GT vectors with improved safety and efficiency properties. Therefore, in collaboration with Dr. Manuel Grez (Georg Speyer Haus, Frankfurt, Germany), we developed next generation γ-retroviral SIN vectors as candidates for future X-CGD GT trials (*chapter 3*). These vectors contain a deletion in the U3 region of the 3' viral LTR, resulting in the inactivation of the viral promoter/enhancer elements after reverse transcription [22,23]. Transgene expression is driven by internal tissue-specific human promoters, which lack enhancer activity and should not be prone to silencing. Among the myelospecific vectors tested, we identified the MRP8 [24] and miRNA-223 [25] promoters to be the most efficient *in vitro* to drive gp91^{phox} expression when used at MOI 1 in human HSC. The X-CGD PLB cell line model proved to be inefficient for testing these vectors, as PLB cells did not form NETs (see objective 1). Only transduced and differentiated human Lin⁻ HSC could form NETs *in vitro* upon stimulation.

In conclusion of objective 3

Our study has lead to identification of pSER11M8delN91 and pSER11miR223delN91 as promising candidates for the next clinical X-CGD GT phase I/II trial, conferring sufficient transgene expression and functional correction of the NADPH oxidase defect with regards to anti-infectious NET-formation. For authorization purposes (Swissmedic), genotoxicity and silencing studies are currently under way, so that the next vector for clinical use will hopefully be available by 2012.

Outlook concerning objective 3

The goal of GT in CGD is to achieve a stable population of gene corrected myeloid precursors, which give rise to a biologically significant number of functional peripheral phagocytes. In order to better evaluate the myelospecificity of above tested vectors and their efficacy in reconstituting antimicrobial defense against pathogens, *in vivo* studies need to be performed in an X-CGD mouse GT model, which will provide more precise information about transgene expression in different blood cell subsets.

Prospects and Perspectives

In addition to the safety issues which are currently addressed in vector development for CGD GT, successful GT for CGD is also limited by the transduction efficiency of myeloid precursors *ex vivo* and by the ability of corrected cells to replace uncorrected cells *in vivo*.

Concerning efficiency, the use of an *in vivo* selectable marker may allow sustained persistence of gene-modified cells at therapeutic relevant levels, although the long-term implications of cytotoxic drug treatment remains a safety concern [26]. To prevent loss of transgene expression, the ubiquitously-acting chromatin opening element (UCOE) has been shown to resist silencing and produce persistent transgene expression in a mouse study [27]. In addition, chromatin insulators such as the chicken β-globin hypersensitive site 4 have been used successfully to shield the surrounding chromatin from the vector and vice versa the vector from influences by the surrounding chromatin, resulting in less position-dependent silencing and in less position-effect variegation [28]. In addition, lentiviral vectors are now being developed and tested for a variety of primary immunodeficiencies, encouraged by the recent first clinical success achieved in X-linked adrenoleukodystrophy (ALD) where a SIN-lentiviral vector in combination with conventional myeloablative conditioning were used [29]. Cartier et al. reported polyclonal reconstitution up to 24 to 30 months from GT, with 9 to 15% granulocytes, monocytes, T and B lymphocytes expressing the ALD protein. The advantage of lentiviral vectors over γ-retroviral vectors is their higher potential to transduce non-dividing stem/progenitor cells avoiding the need for strongly stimulating cell proliferation by cytokines during *ex vivo* culture. This allows for shorter transduction protocols reducing the risk of graft failure and assembly of chromosomal aberrations during prolonged *ex vivo* culture. According to mouse and large animal studies the integration pattern of lentiviral vectors is favorable (compared to γ-retroviral vectors) with a potentially lower risk of insertional mutagenesis [30-32].

Concerning safety, alternatively to the above mentioned approaches with myelospecific SIN vectors, gene correction may be achieved by taking advantage of cellular homologous recombination (HR), one of the major mechanisms for repairing DNA damage, which occur normally during chromosome replication. Zinc finger nucleases are synthetic fusion proteins that combine multiple zinc finger motifs, each capable of binding to specific DNA sequences, with a nuclease that can introduce a DNA break [33]. They might provide integration of the transgene into a so called "safe genomic harbor" (such as the AAVS1 locus on human chromosome 19 [34]) [35,36]. Even if very promising, this strategy is still far from clinical application [37]. The efficacy and potential genotoxicity of such gene correction method have to be well studied but, if safe and effective, this approach could in the future replace gene addition approaches for GT of CGD.

In conclusion, despite considerable progress in the past years, clinical GT for genetic diseases such as CGD remains a field in its infancy, yet/and is opening up fascinating research perspectives.

References

1. Liese J, Kloos S, Jendrossek V, et al. Long-term follow-up and outcome of 39 patients with chronic granulomatous disease. J Pediatr. 2000;137:687-693.
2. Martire B, Rondelli R, Soresina A, et al. Clinical features, long-term follow-up and outcome of a large cohort of patients with Chronic Granulomatous Disease: an Italian multicenter study. Clin Immunol. 2008;126:155-164.
3. Winkelstein JA, Marino MC, Johnston RB, Jr., et al. Chronic granulomatous disease. Report on a national registry of 368 patients. Medicine (Baltimore). 2000;79:155-169.
4. Segal BH, DeCarlo ES, Kwon-Chung KJ, Malech HL, Gallin JI, Holland SM. Aspergillus nidulans infection in chronic granulomatous disease. Medicine (Baltimore). 1998;77:345-354.
5. Seger RA. Modern management of chronic granulomatous disease. Br J Haematol. 2008;140:255-266.
6. Segal BH, Romani LR. Invasive aspergillosis in chronic granulomatous disease. Med Mycol. 2009;47 Suppl 1:S282-290.
7. Bonnett CR, Cornish EJ, Harmsen AG, Burritt JB. Early neutrophil recruitment and aggregation in the murine lung inhibit germination of Aspergillus fumigatus Conidia. Infect Immun. 2006;74:6528-6539.
8. Diamond RD, Clark RA. Damage to Aspergillus fumigatus and Rhizopus oryzae hyphae by oxidative and nonoxidative microbicidal products of human neutrophils in vitro. Infect Immun. 1982;38:487-495.
9. Latge JP. Aspergillus fumigatus and aspergillosis. Clin Microbiol Rev. 1999;12:310-350.
10. Morgenstern DE, Gifford MA, Li LL, Doerschuk CM, Dinauer MC. Absence of respiratory burst in X-linked chronic granulomatous disease mice leads to abnormalities in both host defense and inflammatory response to Aspergillus fumigatus. J Exp Med. 1997;185:207-218.
11. Rex JH, Bennett JE, Gallin JI, Malech HL, Melnick DA. Normal and deficient neutrophils can cooperate to damage Aspergillus fumigatus hyphae. J Infect Dis. 1990;162:523-528.
12. Schaffner A, Douglas H, Braude A. Selective protection against conidia by mononuclear and against mycelia by polymorphonuclear phagocytes in resistance to Aspergillus. Observations on these two lines of defense in vivo and in vitro with human and mouse phagocytes. J Clin Invest. 1982;69:617-631.
13. Zarember KA, Sugui JA, Chang YC, Kwon-Chung KJ, Gallin JI. Human polymorphonuclear leukocytes inhibit Aspergillus fumigatus conidial growth by lactoferrin-mediated iron depletion. J Immunol. 2007;178:6367-6373.

14. Bianchi M, Hakkim A, Brinkmann V, et al. Restoration of NET formation by gene therapy in CGD controls aspergillosis. Blood. 2009;114:2619-2622.
15. Clohessy PA, Golden BE. Calprotectin-mediated zinc chelation as a biostatic mechanism in host defence. Scand J Immunol. 1995;42:551-556.
16. Bianchi M, Niemiec MJ, Siler U, Urban CF, Reichenbach J. Restoration of anti-Aspergillus defense by neutrophil extracellular traps in human chronic granulomatous disease after gene therapy is calprotectin-dependent. J Allergy Clin Immunol. 2011;127:1243-1252 e1247.
17. Dinauer MC, Lekstrom-Himes JA, Dale DC. Inherited Neutrophil Disorders: Molecular Basis and New Therapies. Hematology Am Soc Hematol Educ Program. 2000:303-318.
18. Goldblatt D, Thrasher AJ. Chronic granulomatous disease. Clin Exp Immunol. 2000;122:1-9.
19. Segal BH, Leto TL, Gallin JI, Malech HL, Holland SM. Genetic, biochemical, and clinical features of chronic granulomatous disease. Medicine (Baltimore). 2000;79:170-200.
20. Ott MG, Schmidt M, Schwarzwaelder K, et al. Correction of X-linked chronic granulomatous disease by gene therapy, augmented by insertional activation of MDS1-EVI1, PRDM16 or SETBP1. Nat Med. 2006;12:401-409.
21. Stein S, Ott MG, Schultze-Strasser S, et al. Genomic instability and myelodysplasia with monosomy 7 consequent to EVI1 activation after gene therapy for chronic granulomatous disease. Nat Med. 2010;16:198-204.
22. Kraunus J, Schaumann DH, Meyer J, et al. Self-inactivating retroviral vectors with improved RNA processing. Gene Ther. 2004;11:1568-1578.
23. Schambach A, Mueller D, Galla M, et al. Overcoming promoter competition in packaging cells improves production of self-inactivating retroviral vectors. Gene Ther. 2006;13:1524-1533.
24. Kuwayama A, Kuruto R, Horie N, Takeishi K, Nozawa R. Appearance of nuclear factors that interact with genes for myeloid calcium binding proteins (MRP-8 and MRP-14) in differentiated HL-60 cells. Blood. 1993;81:3116-3121.
25. Johnnidis JB, Harris MH, Wheeler RT, et al. Regulation of progenitor cell proliferation and granulocyte function by microRNA-223. Nature. 2008;451:1125-1129.
26. Beard BC, Trobridge GD, Ironside C, McCune JS, Adair JE, Kiem HP. Efficient and stable MGMT-mediated selection of long-term repopulating stem cells in nonhuman primates. J Clin Invest;120:2345-2354.

27. Zhang F, Thornhill SI, Howe SJ, et al. Lentiviral vectors containing an enhancer-less ubiquitously acting chromatin opening element (UCOE) provide highly reproducible and stable transgene expression in hematopoietic cells. Blood. 2007;110:1448-1457.
28. Aker M, Tubb J, Groth AC, et al. Extended core sequences from the cHS4 insulator are necessary for protecting retroviral vectors from silencing position effects. Hum Gene Ther. 2007;18:333-343.
29. Cartier N, Hacein-Bey-Abina S, Bartholomae CC, et al. Hematopoietic stem cell gene therapy with a lentiviral vector in X-linked adrenoleukodystrophy. Science. 2009;326:818-823.
30. Hematti P, Hong BK, Ferguson C, et al. Distinct genomic integration of MLV and SIV vectors in primate hematopoietic stem and progenitor cells. PLoS Biol. 2004;2:e423.
31. Montini E, Cesana D, Schmidt M, et al. Hematopoietic stem cell gene transfer in a tumor-prone mouse model uncovers low genotoxicity of lentiviral vector integration. Nat Biotechnol. 2006;24:687-696.
32. Montini E, Cesana D, Schmidt M, et al. The genotoxic potential of retroviral vectors is strongly modulated by vector design and integration site selection in a mouse model of HSC gene therapy. J Clin Invest. 2009;119:964-975.
33. Kim YG, Cha J, Chandrasegaran S. Hybrid restriction enzymes: zinc finger fusions to Fok I cleavage domain. Proc Natl Acad Sci U S A. 1996;93:1156-1160.
34. Dutheil N, Shi F, Dupressoir T, Linden RM. Adeno-associated virus site-specifically integrates into a muscle-specific DNA region. Proc Natl Acad Sci U S A. 2000;97:4862-4866.
35. Bohne J, Cathomen T. Genotoxicity in gene therapy: an account of vector integration and designer nucleases. Curr Opin Mol Ther. 2008;10:214-223.
36. Cathomen T, Segal DJ, Brondani V, Muller-Lerch F. Generation and functional analysis of zinc finger nucleases. Methods Mol Biol. 2008;434:277-290.
37. Lombardo A, Genovese P, Beausejour CM, et al. Gene editing in human stem cells using zinc finger nucleases and integrase-defective lentiviral vector delivery. Nat Biotechnol. 2007;25:1298-1306.

Abbreviations

AAV	adeno-associated virus
ADA-SCID	adenosine deiminase-SCID
ALD	adrenoleukodystrophy
BM	bone marrow
BMT	bone marrow transplantation
CFU	colony forming unit
CGD	chronic granulomatous disease
DAPI	4',6-diamidino-2-phenylindole
DHR	dihydrorhodamine
DMF	dimethylformamide
DNA	deoxyribonucleic acid
DNase	deoxyribonuclease
ds	double stranded
ELISA	enzyme-linked immunosorbent assay
Env	envelope gene
Evi1	ecotropic viral integration 1 site
FACS	fluorescence-activated cell sorter
FAD	flavin adenin dinucleotide
FBS	fetal bovine serum
FDG	fluorodeoxyglucose
FITC	fluorescin isothiocyanate
Flt3L	Fms-related tyrosine kinase 3 ligand
Gag	group-specific antigen gene
GALV	gibbon ape leukemia virus
G-CSF	granulocyte-colony stimulating factor
GDP	guanosine diphosphate
GFP	green fluorescent protein
GT	gene therapy
GTP	guanosine triphosphate
GVHD	graft-versus-host disease
H_2O_2	hydrogen peroxide
HEK	human embryonic kidney
HGMD	Human Gene Mutation Database
HIV	human immunodeficiency virus
HLA	human leukocyte antigen
HOBr	hypobromous acid

Abbreviations

HOCl	hypochlorous acid
HR	homologous recombination
HSC	hematopoietic stem cell
IDO	indoleamine 2,3-dioxygenase
IFN-γ	interferon-gamma
IL	interleukin
Lin⁻	lineage negative cells
LMO2	LIM domain only 2
LRR	leucine-rich repeat
LTR	long terminal repeat
MDS1	myelodysplastic syndrome 1
MFI	mean fluorescence intensity
miRNA-223	microRNA-223
MLV	murine leukemia virus
MMP	matrix metalloproteinase
MNase	micrococcal nuclease
MOI	multiplicity of infection
MPO	myeloperoxidase
MRP8	migration inhibitory factor-related protein 8
MyD88	myeloid differentiation primary-response protein 88
NADPH	reduced nicotinamide adenine dinucleotide phosphate
NBT	nitroblue tetrazolium
NE	neutrophil elastase
NET	neutrophil extracellular trap
NOX	NADPH oxidase
O_2^-	superoxide
OH^-	hydroxyl radicals
ORF	open reading frame
PAD4	peptidylarginine deiminase
pDC	plasmacytoid dendritic cell
PET-CT	positron emission tomography-computed tomography
PHOX	phagocyte oxidase
PI3K	phosphatidylinositol 3-kinase
PMA	phorbol 12-myristate 13-acetate
Pol	polymerase gene
PPT	polypurine-tract
PRDM	PR domain

Abbreviations

Pro	protease gene
Psi (ψ)	packaging signal
PU.1	purine rich box-1
RNA	ribonucleic acid
ROS	reactive oxygen species
SCF	stem cell factor
SCID	severe combined immunodeficiency
SD	standard deviation
SEM	scanning electron microscopy
SFFV	spleen focus-forming virus
SH	Src homology
SIN	self-inactivating
SLE	systemic lupus erythematosis
SOD	superoxide dismutase
SP107	synthetic promoter 107
ss	single stranded
TPO	thrombopoietin
TU	transducing unit
UCOE	ubiquitously-acting chromatin opening element
VSV	vesicular stomatitis virus
WPRE	woodchuck hepatitis virus post transcriptional regulatory element
X-CGD	X-linked CGD
XTT	2,3-bis(2-Methoxy-4-nitro-5-sulfophenyl)-2H-tetrazolium-5-carboxanilide

ORIGINAL PUBLICATIONS

GENE THERAPY

Brief report

Restoration of NET formation by gene therapy in CGD controls aspergillosis

*Matteo Bianchi,[1] *Abdul Hakkim,[2] Volker Brinkmann,[3] Ulrich Siler,[1] Reinhard A. Seger,[1] †Arturo Zychlinsky,[2] and †Janine Reichenbach[1]

[1]Division of Immunology/Haematology/BMT, University Children's Hospital Zurich, Zurich, Switzerland; and [2]Department of Cellular Microbiology and [3]Microscopy Core Facility, Max Planck Institute for Infection Biology, Berlin, Germany

Chronic granulomatous disease (CGD) patients have impaired nicotinamide adenine dinucleotide phosphate (NADPH) oxidase function, resulting in poor antimicrobial activity of neutrophils, including the inability to generate neutrophil extracellular traps (NETs). Invasive aspergillosis is the leading cause of death in patients with CGD; it is unclear how neutrophils control *Aspergillus* species in healthy persons. The aim of this study was to determine whether gene therapy restores NET formation in CGD by complementation of NADPH oxidase function, and whether NETs have antimicrobial activity against *Aspergillus nidulans*. Here we show that reconstitution of NET formation by gene therapy in a patient with CGD restores neutrophil elimination of *A nidulans* conidia and hyphae and is associated with rapid cure of preexisting therapy refractory invasive pulmonary aspergillosis, underlining the role of functional NADPH oxidase in NET formation and antifungal activity. (Blood. 2009;114: 2619-2622)

Introduction

Activated neutrophils kill microbes intracellularly after phagocytosis and by extracellular mechanisms, including neutrophil extracellular traps (NETs), which are composed of chromatin decorated with granular proteins.[1] NETs bind bacteria[1] and fungi[2] and expose antimicrobial molecules. Generation of NETs requires reactive oxygen species produced by the nicotinamide adenine dinucleotide phosphate (NADPH) oxidase.[3]

Chronic granulomatous disease (CGD) is caused by mutations in genes encoding NADPH oxidase subunits. CGD patients do not produce reactive oxygen species, kill microbes poorly, and are susceptible to recurrent life-threatening infections.[4] *Aspergillus spp* infections cause pneumonia and disseminated disease and are the leading cause of death in these patients.[4-6]

It is unclear how *Aspergillus* infections are controlled in healthy persons.[7-13] In CGD patients, these infections are frequently refractory to antifungal therapy, treatment with interferon-γ, or granulocyte transfusions.[5] Here we show that the recently discovered NADPH oxidase-dependent microbicidal pathway through NETs[1-3] is efficient against *Aspergillus nidulans* conidia and hyphae in vitro and that restoration of NET formation by GT of X-CGD aided clearing severe invasive *A nidulans* infection in vivo.

Methods

Gene therapy

We treated an 8.5-year-old boy with X-linked gp91^{phox}-deficient CGD and therapy refractory *A nidulans* lung infection with a monocistronic long terminal repeat-driven gamma-retroviral SF71gp91^{phox} vector (see supplemental data, available on the *Blood* website; see the Supplemental Materials link at the top of the online article). The protocol for the patient's treatment was approved by the ethics review board of the University Children's Hospital Zurich and the Swiss Expert Committee for Bio-Safety, after written informed consent from his parents in accordance with the Declaration of Helsinki. For follow-up monitoring, gp91^{phox} expression was measured by fluorescence-activated cell sorter (FACS) on peripheral neutrophils after 30 minutes of staining at room temperature with 10 μg/mL gp91^{phox}-fluorescein isothiocyanate (FITC) antibody (Anti-Flavocytochrome b558, clone 7D5, MBL). NADPH oxidase activity was measured by standard dihydrorhodamine and nitroblue tetrazolium tests (supplemental materials). Bone marrow colony assays and determination of proviral gp91^{phox} sequences in genomic DNA were performed as described.[14]

NET induction

NET formation was visualized as described (supplemental materials) and quantified after stimulation of 5×10^4 neutrophils for 3 hours with 40 nM phorbol 12-myristate 13-acetate (PMA) and staining the NET-DNA with 1 μM Sytox green (Invitrogen) in a black 96-well plate (BD Biosciences). The plates were read in a fluorescence microplate reader (Victor,[3] PerkinElmer Life and Analytical Sciences) with a filter setting of 485 nm/535 nm (excitation/emission).

NET antifungal activity

The *A nidulans* strain used was isolated from bronchoalveolar lavage fluid of the patient; conidia were grown and collected as described.[7] Neutrophils after GT were stained with gp91^{phox}-FITC antibody and sorted by FACS (FACSAria, BD Biosciences) into gp91^{phox}-negative (gp91^{phox-}) and -positive (gp91^{phox+}) populations. A total of 10^5 neutrophils were activated with PMA (40 nM) at 37°C for 4 hours in a 96-well plate, and then infected with conidia (multiplicity of infection *A nidulans*/neutrophils = 0.5) plus or minus prior digestion of NETs with 10 U/mL of Micrococcal Nuclease (MNase, Worthington Biochemical) for 30 minutes. Afterward the plates were centrifuged for 5 minutes at 400g and incubated for 16 hours at 37°C,

---

Submitted May 12, 2009; accepted June 8, 2009. Prepublished online as *Blood* First Edition paper, June 18, 2009; DOI 10.1182/blood-2009-05-221606.

*M.B. and A.H. contributed equally to this study.

†A.Z. and J.R. contributed equally to this study.

The online version of this article contains a data supplement.

The publication costs of this article were defrayed in part by page charge payment. Therefore, and solely to indicate this fact, this article is hereby marked "advertisement" in accordance with 18 USC section 1734.

© 2009 by The American Society of Hematology

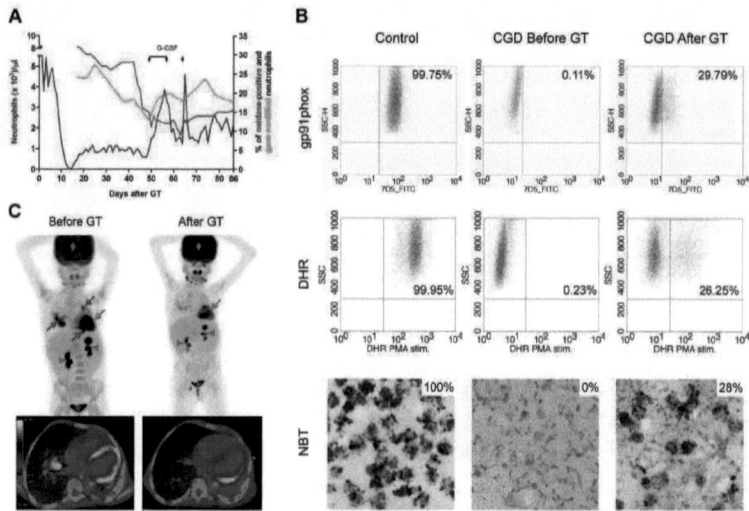

Figure 1. Restoration of NADPH oxidase function. (A) Hematopoietic reconstitution and gene marking after GT. Absolute neutrophil counts (left y-axis), quantification of gene-modified cells in peripheral neutrophils by quantitative polymerase chain reaction and quantification of neutrophils with NADPH oxidase activity by DHR test (right y-axis) are shown. When the percentage of transduced neutrophils decreased, granulocyte colony-stimulating factor (5 μg/kg per day subcutaneously) was administered on days 49 to 57 and on day 64. (B) Reconstitution of NADPH oxidase activity. Before and 6 weeks after GT, gp91phox protein expression was measured by FACS analysis after 30 minutes of staining with 10 μg/mL gp91phox-FITC antibody. Superoxide production was assessed by oxidation of DHR on stimulation with PMA and by reduction of nitroblue tetrazolium to formazan (dark precipitate) after stimulation with opsonized zymosan. The thresholds were determined using unstained (FACS) or unstimulated (DHR) cells for each experiment. (C) PET-CT scan. Before GT, PET-CT scan showed several active infectious foci with fluorine-18-fluoro-2-deoxy-D-glucose uptake in both lungs of the patient (red arrows); the infection cleared 6 weeks after administration of gene-corrected cells. In green, physiologic FDG uptake in heart (arrow), kidneys (arrowheads), bladder (*), and brain (diamond) are indicated for reference.

allowing germination and hyphal outgrowth. Alternatively, 5×10^4 conidia were incubated for 12 hours at 37°C to allow hyphal outgrowth; then 10^5 neutrophils were added, centrifuged for 5 minutes at 400g, and incubated for 2 and 5 hours with PMA (40 nM) to induce NET formation, plus or minus 10 U/mL MNase. Fungal growth was quantified with XTT (Invitrogen) as described.[15]

Results and discussion

In healthy subjects, NETs might be essential to eliminate fungi because hyphae are too large to be phagocytosed.[7,9,12,13,16-21] CGD patients are unable to make NETs.[3] Indeed, neutrophils of our X-CGD patient could only make NETs and inhibit growth of *A nidulans* after genetic complementation by GT. Neutrophils expressing functional gp91phox increased from 0% to 26% to 29% 6 weeks after GT (Figure 1A-B), then decreased and leveled at approximately 16% for up to 3 months. The *A nidulans* infection completely cleared 6 weeks after GT (Figure 1C), correlating with the rise in neutrophils with NADPH oxidase activity.

The patient's neutrophils did not make NETs before GT as analyzed by fluorescence (not shown), immunofluorescence, and scanning electron microscopy (Figure 2A-B, D-E).[3] After GT, the patient's neutrophils made NETs (Figure 2C,F), the percentage of cells releasing NET-DNA (28%; Figure 2G) correlating with the level of oxidase chimerism (Figure 1B). Activation of sorted neutrophils showed that reconstitution with functional NADPH oxidase allowed corrected CGD neutrophils to make NETs (Figure 2H).

To test whether efficient eradication of the patient's infection was the result of the recovered ability to make NETs, neutrophils were infected with the *A nidulans* strain isolated from the patient. Approximately 80% of conidia germination (Figure 2I) and 45% of hyphal growth (Figure 2J) were inhibited by CGD gp91phox+ neutrophils, comparable with the antimicrobial activity of control neutrophils. CGD gp91phox− neutrophils were inefficient in controlling fungal growth (Figure 2I-J, supplemental Figure 1). When NETs were dismantled with MNase before infection, fungicidal activity was abrogated to that of CGD gp91phox− neutrophils.

In the absence of NETs, control of hyphal growth was independent of NADPH oxidase activity: when neutrophils were infected after 2-hour PMA stimulation, before NETs had been made, CGD gp91phox+, CGD gp91phox−, and neutrophils from healthy donor controlled growth of *A nidulans* with similar modest efficiency (Figure S2) in a NET-independent fashion because antimicrobial activity was not affected by MNase. This limited NET-independent antimicrobial activity, presumably by conidia phagocytosis, degranulation, or unknown mechanisms, suggests an NADPH oxidase-independent antifungal mechanism, clinically ineffective before GT. These data propose that the patient's clearance of fungal infection after GT was controlled by NETs. Definitive in vivo proof is obviously technically impossible.

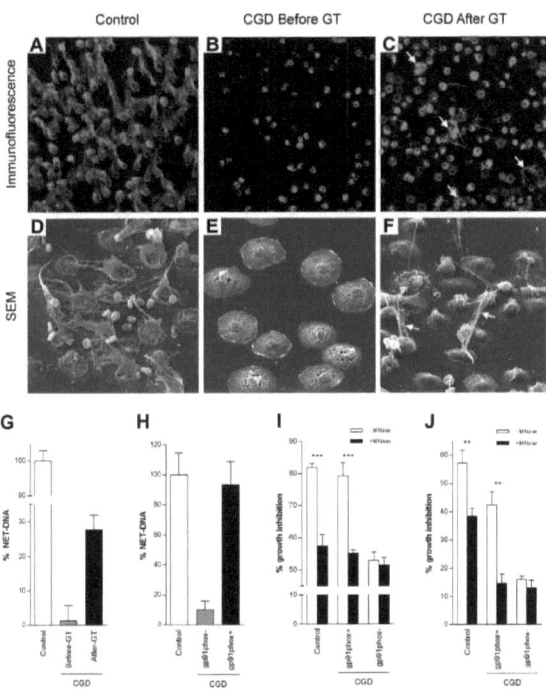

Figure 2. NET formation and inhibition of *A nidulans* growth. Control (A,D), but not CGD (B,E), neutrophils made NETs on 3-hour PMA stimulation. For immunofluorescence, NETs were stained with an antibody that recognizes neutrophil elastase (green; A-C). NETs were clearly visible also by scanning electron microscopy (SEM; D-F). Neutrophils isolated from the CGD patient before GT could be activated because they flattened out (E) but did not make NETs. The ability to form NETs was partially restored by GT 6 weeks after GT (C,F white arrows). (G) Quantification of NET-DNA released after 3 hours of PMA stimulation of control neutrophils, CGD neutrophils before and 6 weeks after GT or (H) after stimulation of CGD gp91^{phox+} and CGD gp91^{phox-} FACS-sorted neutrophils. CGD gp91^{phox+} neutrophils showed normal NET formation, whereas CGD gp91^{phox-} neutrophils showed only residual NET formation. FACS-sorting efficiency was 90% to 92% for CGD gp91^{phox-} and 95% to 96% for CGD gp91^{phox+} cells. (I-J) NET inhibition of *A nidulans* conidia and hyphae. (I) Conidia were plated on FACS-sorted neutrophils prestimulated with PMA plus or minus MNAse (ie, when NET formation was complete, cells were dead and therefore incapable of phagocytosis). Hyphal outgrowth was measured after 16 hours. (J) Hyphae were coincubated with FACS-sorted neutrophils, and PMA plus or minus MNAse and hyphal viability was assessed after 5 hours. (G-J) Data are mean ± SD of a representative triplicate experiment. Inhibition of fungal growth is expressed as percentage of control values (*A nidulans* conidia or hyphae incubated in media). The differences between -MNase and +MNase were significant (for control and CGD gp91^{phox+} cells) by Student *t* test: **$P < .01$; ***$P < .001$.

Alveolar macrophages probably constitute the first line of defense to conidia that escape mucociliary clearance in healthy persons.[13,16] Whether reconstituted NADPH oxidase function in alveolar macrophages also contributed to microbial killing in the patient presented is difficult to assess. In neutrophils, however, conidia resist intracellular killing because of their relative tolerance to reactive oxygen species.[7,9,12,19-22] Our results suggest that conidia are killed mainly extracellularly rather than after phagocytosis. Both conidia and hyphae get ensnared by neutrophils and probably killed within NETs by concentrated antimicrobials. Cooperation of gp91^{phox-} and gp91^{phox+} neutrophils in NET antifungal activity is doubtful because we showed that gp91^{phox-} neutrophils do not make NETs when coincubated with gp91^{phox+} neutrophils (unsorted cells in Figure 2F-G),[10] suggesting that the amount of H_2O_2 released by gp91^{phox+} neutrophils is insufficient to induce NETs in gp91^{phox-} cells.[3]

A GT approach to treat CGD may be used to overcome recalcitrant, life-threatening infections but is currently limited as salvage therapy to experimental studies in selected patients with very poor performance status and lacking an human leukocyte antigen-identical hematopoietic stem cell donor.[4] GT was rapidly beneficial to our CGD patient who had suffered from an otherwise incurable fungal infection. Until day 86 after GT, there was no clonal dominance in bone marrow culture-derived CD34+ cells (not shown) or expansion of gene-corrected cells in blood (Figure 1A). There is a risk, however, of insertional mutagenesis by transactivating retroviral vector insertions into proto-oncogenes, as shown in a recent GT trial with 2 adult CGD patients who developed monosomy 7 and myelodysplastic syndrome (M. Grez, Institute of Biomedical Research, Georg-Speyer-Haus, Frankfurt, Germany, oral communication, April 2009) using the same gamma-retroviral SF71gp91^{phox} vector.[14] In addition, 5 patients developed leukemia in 2 GT trials in children with severe combined immunodeficiency.[23,24] These experiences mandate the careful follow-up of patients.

In conclusion, we show that the severe immunodeficient phenotype and the high susceptibility to *Aspergillus* infection of CGD patients might be linked to the absence of NETs, and that restoration of NADPH oxidase function and NET formation by GT leads to rapid cure of refractory invasive aspergillosis in X-linked CGD.

Acknowledgments

The authors thank the patient and his family for their trust; the medical and nursing staff of the bone marrow transplantation unit of University Children's Hospital Zurich; Manuel Grez and Klaus Kühlke for developing and providing the SF71gp91^{phox} vector, respectively; Maja Rutishauser, Corinne Wenk, and Oralea Büchi for technical assistance; Ursula Lüthi and Klaus Marquardt for electron microscopy; Britta Laube for help with immunofluorescence imaging; Alex Imhof for isolating *A nidulans* conidia; and Hans Steinert for carrying out PET-CT scans.

This work was supported by a grant of the Chronic Granulomatous Disorder Research Trust, United Kingdom (J.R., M.B.), a Forschungskredit der Universität Zürich 2006 grant (J.R., M.B.), and a grant from the Stiftung für wissenschaftliche Forschung an der Universität Zürich/Baugarten Stiftung (R.S.). The funders had no role in study design, data collection and analysis, decision to publish, or preparation of the manuscript.

Authorship

Contribution: M.B. and A.H. performed the experiments, analyzed data, and contributed to the writing of the manuscript; V.B. did the immunofluorescence image acquisition and contributed to data analysis; U.S. did the bone marrow cultures and contributed to data analysis; R.A.S. designed the clinical gene therapy protocol, attended the patient together with J.R., and contributed to writing of the manuscript; and A.Z. and J.R. designed and directed the study and contributed to the writing of the manuscript.

Conflict-of-interest disclosure: The authors declare no competing financial interests.

Correspondence: Janine Reichenbach, Division of Immunology/Haematology/BMT, University Children's Hospital Zurich, Steinwiesstrasse 75, 8032 Zurich, Switzerland; e-mail: janine.reichenbach@kispi.uzh.ch; or Arturo Zychlinsky, Department of Cellular Microbiology, Max Planck Institute for Infection Biology, Charitéplatz 1, Berlin 10117, Germany; e-mail: zychlinsky@mpiib-berlin.mpg.de.

References

1. Brinkmann V, Reichard U, Goosmann C, et al. Neutrophil extracellular traps kill bacteria. *Science*. 2004;303:1532-1535.
2. Urban CF, Reichard U, Brinkmann V, Zychlinsky A. Neutrophil extracellular traps capture and kill Candida albicans yeast and hyphal forms. *Cell Microbiol*. 2006;8:668-676.
3. Fuchs TA, Abed U, Goosmann C, et al. Novel cell death program leads to neutrophil extracellular traps. *J Cell Biol*. 2007;176:231-241.
4. Seger RA. Modern management of chronic granulomatous disease. *Br J Haematol*. 2008;140:255-266.
5. Segal BH, DeCarlo ES, Kwon-Chung KJ, Malech HL, Gallin JI, Holland SM. Aspergillus nidulans infection in chronic granulomatous disease. *Medicine (Baltimore)*. 1998;77:345-354.
6. Winkelstein JA, Marino MC, Johnston RB Jr, et al. Chronic granulomatous disease: report on a national registry of 368 patients. *Medicine (Baltimore)*. 2000;79:155-169.
7. Bonnett CR, Cornish EJ, Harmsen AG, Burritt JB. Early neutrophil recruitment and aggregation in the murine lung inhibit germination of Aspergillus fumigatus Conidia. *Infect Immun*. 2006;74:6528-6539.
8. Diamond RD, Clark RA. Damage to Aspergillus fumigatus and Rhizopus oryzae hyphae by oxidative and nonoxidative microbicidal products of human neutrophils in vitro. *Infect Immun*. 1982;38:487-495.
9. Morgenstern DE, Gifford MA, Li LL, Doerschuk CM, Dinauer MC. Absence of respiratory burst in X-linked chronic granulomatous disease mice leads to abnormalities in both host defense and inflammatory response to Aspergillus fumigatus. *J Exp Med*. 1997;185:207-218.
10. Rex JH, Bennett JE, Gallin JI, Malech HL, Melnick DA. Normal and deficient neutrophils can cooperate to damage Aspergillus fumigatus hyphae. *J Infect Dis*. 1990;162:523-528.
11. Schaffner A, Douglas H, Braude A. Selective protection against conidia by mononuclear and against mycelia by polymorphonuclear phagocytes in resistance to Aspergillus: observations on these two lines of defense in vivo and in vitro with human and mouse phagocytes. *J Clin Invest*. 1982;69:617-631.
12. Zarember KA, Sugui JA, Chang YC, Kwon-Chung KJ, Gallin JI. Human polymorphonuclear leukocytes inhibit Aspergillus fumigatus conidial growth by lactoferrin-mediated iron depletion. *J Immunol*. 2007;178:6367-6373.
13. Latgé JP. Aspergillus fumigatus and aspergillosis. *Clin Microbiol Rev*. 1999;12:310-350.
14. Ott MG, Schmidt M, Schwarzwaelder K, et al. Correction of X-linked chronic granulomatous disease by gene therapy, augmented by insertional activation of MDS1-EVI1, PRDM16 or SETBP1. *Nat Med*. 2006;12:401-409.
15. Meshulam T, Levitz SM, Christin L, Diamond RD. A simplified new assay for assessment of fungal cell damage with the tetrazolium dye, (2,3)-bis-(2-methoxy-4-nitro-5-sulphenyl)-(2H)-tetrazolium-5-carboxanilide (XTT). *J Infect Dis*. 1995;172:1153-1156.
16. Mizgerd JP. Acute lower respiratory tract infection. *N Engl J Med*. 2008;358:716-727.
17. Ibrahim-Granet O, Philippe B, Boleti H, et al. Phagocytosis and intracellular fate of Aspergillus fumigatus conidia in alveolar macrophages. *Infect Immun*. 2003;71:891-903.
18. Philippe B, Ibrahim-Granet O, Prevost MC, et al. Killing of Aspergillus fumigatus by alveolar macrophages is mediated by reactive oxidant intermediates. *Infect Immun*. 2003;71:3034-3042.
19. Cornish EJ, Hurtgen BJ, McInnerney K, et al. Reduced nicotinamide adenine dinucleotide phosphate oxidase-independent resistance to Aspergillus fumigatus in alveolar macrophages. *J Immunol*. 2008;180:6854-6867.
20. Levitz SM, Farrell TP. Human neutrophil degranulation stimulated by Aspergillus fumigatus. *J Leukoc Biol*. 1990;47:170-175.
21. Segal AW. How neutrophils kill microbes. *Annu Rev Immunol*. 2005;23:197-223.
22. Lehrer RI, Jan RG. Interaction of Aspergillus fumigatus spores with human leukocytes and serum. *Infect Immun*. 1970;1:345-350.
23. Hacein-Bey-Abina S, Garrigue A, Wang GP, et al. Insertional oncogenesis in 4 patients after retrovirus-mediated gene therapy of SCID-X1. *J Clin Invest*. 2008;118:3132-3142.
24. Howe SJ, Mansour MR, Schwarzwaelder K, et al. Insertional mutagenesis combined with acquired somatic mutations causes leukemogenesis following gene therapy of SCID-X1 patients. *J Clin Invest*. 2008;118:3143-3150.

Response

Protecting against *Aspergillus* infection in CGD

We would like to thank Remijsen and colleagues for their comments regarding our recent publication in *Blood* titled "Restoration of NET formation by gene therapy in CGD controls aspergillosis."[1]

We regret that we neglected to cite the work of Romani et al.[2] This was because the "Brief Report" format in *Blood* is restricted in length and number of citations. Romani et al reported that superoxide produced by the nicotinamide adenine dinucleotide phosphate (NADPH) oxidase regulates indoleamine 2,3-dioxygenase (IDO), which leads to L-kynurenine formation and eventually to the activation of interleukin-17 (IL-17)–producing T cells, which are involved in the acquired immune response. Indeed, treatment with L-kynurenine and interferon-γ helps resolve *Aspergillus* infections in p47[phox]-deficient mice, a murine model of chronic granulomatous disease (CGD).

We analyzed the level of L-kynurenine and IL-17 in sera collected from the CGD patient described in our publication[1] before, as well as 11, 17, and 98 days after, gene therapy (GT). There were no significant differences in L-kynurenine levels in the sera before and after GT as measured by high-performance liquid chromatography.[3] The concentration of IL-17 in sera was less than the detection limit of 30 pg/mL of the enzyme-linked immunosorbent assay (R&D Systems) at all time points analyzed. Because of ethical considerations, we did not analyze local L-kynurenine and IL-17 levels in tissue biopsies, and therefore cannot exclude a local effect of the IDO pathway on *Aspergillus* infection.

The complex clinical phenotype in CGD patients reflects the pleiotropic functions of the NADPH oxidase in immune defense. It is likely that several mechanisms, including the activity of IDO, contribute to the restoration of immune defense after GT, and it would certainly be important to address this point if the opportunity arises. Treatment with L-kynurenine, however, has not yet been tested in CGD patients because of concerns about its epileptogenic potential.

The high incidence of aspergillosis compared with other opportunistic infections in CGD patients is indeed difficult to explain. It would certainly be interesting to survey how neutrophils isolated from these patients respond to a spectrum of different opportunistic microbes. Interestingly, knockout mice in any of the NADPH oxidase subunits seem susceptible to diverse microbial insults. Regardless, it is clear that the NADPH oxidase, as a part of the innate immune system, is initially important in microbe clearance, through, for example, phagocytosis and neutrophil extracellular trap (NET) formation. This enzyme, as elegantly shown by Romani et al, is also important in the second function of innate immunity, which is to set the stage for an acquired immune response.

We reported the acquisition of in vitro anti-*Aspergillus* activity of NETs after GT. This is in line with the observations that fungi are more susceptible to NETs than to phagocytic killing.[4] More importantly, the patient started to get better within few days after engraftment of gene-transduced cells, suggesting an immediate innate response. The role of NETs in the recovery of the patient will remain a correlation since we reported the in vitro function of human cells; in vivo experiments are obviously out of the question. It is likely that the improvement after restoration of NADPH oxidase activity in the CGD patient reported was due to several pathways where this enzyme is involved. Our data, however, suggest that NETs played a prominent role in the clearance of the *Aspergillus* infection.

Abdul Hakkim
Department of Cellular Microbiology,
Max Planck Institute for Infection Biology,
Berlin, Germany

Robert Hurwitz
Protein Facility, Max Planck Institute for Infection Biology,
Berlin, Germany

Matteo Bianchi
Division of Immunology/Haematology/BMT,
University Children's Hospital,
Zurich, Switzerland

Volker Brinkmann
Microscopy Core Facilities, Max Planck Institute for Infection Biology,
Berlin, Germany

Ulrich Siler
Division of Immunology/Haematology/BMT,
University Children's Hospital,
Zurich, Switzerland

Reinhard A. Seger
Division of Immunology/Haematology/BMT,
University Children's Hospital,
Zurich, Switzerland

Arturo Zychlinsky
Department of Cellular Microbiology,
Max Planck Institute for Infection Biology,
Berlin, Germany

Janine Reichenbach
Division of Immunology/Haematology/BMT,
University Children's Hospital,
Zurich, Switzerland

Approval was obtained from the ethics review board of the University Children's Hospital Zurich and the Swiss Expert Committee for Bio-Safety for these studies. Informed consent was provided according to the Declaration of Helsinki.

Conflict-of-interest disclosure: The authors declare no competing financial interests.

Correspondence: Janine Reichenbach, University Children's Hospital Zurich, Steinwiesstr 75, Zurich 8032, Switzerland; e-mail: janine.reichenbach@kispi.uzh.ch.

References

1. Bianchi M, Hakkim A, Brinkmann V, et al. Restoration of NET formation by gene therapy in CGD controls aspergillosis. *Blood*. 2009;114:2619-2622.
2. Romani L, Fallarino F, De Luca A, et al. Defective tryptophan catabolism underlies inflammation in mouse chronic granulomatous disease. *Nature*. 2008;451:211-215.
3. Widner B, Werner ER, Schennach H, Wachter H, Fuchs D. Simultaneous measurement of serum tryptophan and kynurenine by HPLC. *Clin Chem*. 1997;43:2424-2426.
4. Urban CF, Reichard U, Brinkmann V, Zychlinsky A. Neutrophil extracellular traps capture and kill Candida albicans yeast and hyphal forms. *Cell Microbiol*. 2006;8:668-676.

Original Publications

Restoration of anti-*Aspergillus* defense by neutrophil extracellular traps in human chronic granulomatous disease after gene therapy is calprotectin-dependent

Matteo Bianchi, MSc,[a]* Maria J. Niemiec, MSc,[b]* Ulrich Siler, PhD,[a] Constantin F. Urban, PhD,[b]‡ and Janine Reichenbach, MD[a]‡   *Zurich, Switzerland, and Umeå, Sweden*

Background: *Aspergillus* spp infection is a potentially lethal disease in patients with neutropenia or impaired neutrophil function. We showed previously that *Aspergillus* hyphae, too large for neutrophil phagocytosis, are inhibited by reactive oxygen species–dependent neutrophil extracellular trap (NET) formation. This process is defective in chronic granulomatous disease (CGD) because of impaired phagocyte nicotinamide adenine dinucleotide phosphate (NADPH) oxidase function.
Objective: To determine the antifungal agent and mechanism responsible for reconstitution of *Aspergillus* growth inhibition within NETs after complementation of NADPH oxidase function by gene therapy (GT) for CGD.
Methods: Antifungal activity of free and NET-released calprotectin was assessed by incubation of *Aspergillus nidulans* with purified calprotectin, induced NETs from human controls, and CGD neutrophils after GT in the presence or absence of Zn^{2+} or α-S100A9 antibody, and with induced NETs from wild-type or $S100A9^{-/-}$ mouse neutrophils.
Results: We identified the host Zn^{2+} chelator calprotectin as a neutrophil-associated antifungal agent expressed within NETs, reversibly preventing *A nidulans* growth at low concentrations, and leading to irreversible fungal starvation at higher concentrations. Specific antibody-blocking and Zn^{2+} addition abolished calprotectin-mediated inhibition of *A nidulans* proliferation *in vitro*. The role of calprotectin in anti-*Aspergillus* defense was confirmed in calprotectin knockout mice.
Conclusion: Reconstituted NET formation by GT for human CGD was associated with rapid cure of pre-existing therapy-refractory invasive pulmonary aspergillosis *in vivo*, underlining the role of functional NADPH oxidase in NET formation and calprotectin release for antifungal activity. These results demonstrate the critical role of calprotectin in human innate immune defense against *Aspergillus* infection. (J Allergy Clin Immunol 2011;127:1243-52.)

Key words: Chronic granulomatous disease, gene therapy, neutrophil extracellular trap, calprotectin, Aspergillus *infection*

Neutrophils kill microbes by distinct reactive oxygen species (ROS)–dependent processes: intracellularly after phagocytosis and by extracellular mechanisms including neutrophil extracellular traps (NETs), which capture and kill bacteria,[1] parasites,[2] and fungi.[3,4] NETs are composed of chromatin (histones and DNA), decorated with at least 20 other proteins.[1,4,5] In naive neutrophils, 8 of these proteins localize to granules, such as neutrophil elastase, myeloperoxidase, or azurocidin, and 11 to the cytoplasm, such as catalase and the calprotectin complex.[4] This heteroduplex formed by subunits S100A8 and S100A9 belongs to the calcium-binding protein family of S100 proteins and exerts strong microbistatic activity against a variety of microorganisms *in vitro*.[6]

Defective nicotinamide adenine dinucleotide phosphate (NADPH) oxidase function accounts for impaired phagocyte ROS production and poor microbial killing in chronic granulomatous disease (CGD), leading to recurrent life-threatening bacterial and fungal infections. *Aspergillus* spp infections cause pneumonia and disseminated disease and are the leading cause of death in these patients.[7-10] ROS production by the NADPH oxidase is indispensable for microbe-triggered NET formation,[5,11,12] whereas their mechanism of action in this process is unknown. After release from granules, neutrophil elastase (NE) and myeloperoxidase synergize to drive decondensation of chromatin, an early event preceding the release of NETs.[13,14] ROS might contribute to the release of NE and myeloperoxidase from granules. We previously reported impaired NET formation in CGD *in vitro*, resulting in defective clearance of *Aspergillus* infection *in vivo*. NADPH oxidase–dependent ROS production and NET inhibition of *Aspergillus* conidia and hyphae were restored early after gene therapy (GT) in a patient with X-linked gp91[phox]-deficient CGD, paralleled by clearance of therapy-refractory *Aspergillus nidulans* lung infection.[11]

The antifungal agent responsible for *A nidulans* growth inhibition within NETs has not been characterized. Here we show that among the 24 NET-associated proteins, the major

From [a]the Division of Immunology/Hematology/BMT, University Children's Hospital Zurich; and [b]the Antifungal Immunity Group, Molecular Biology Department, Laboratory for Molecular Infection Medicine Sweden, Umeå University.
*These authors contributed equally to this work.
‡These authors contributed equally to this work.
Supported by a grant of the Chronic Granulomatous Disorder Research Trust, United Kingdom (grant no. J4G/08/01 to J.R. and M.B.), a grant of GEBERT RÜF STIFTUNG, Switzerland (grant number GRS-046/10 to J.R.), a grant of the EU FP7 CELL-PID programme (J.R. and U.S.), and a grant of the Swiss National Science Foundation (grant no. 320000-121983 to U.S.). M.J.N. and C.F.U. were supported by a starting grant from MIMS, a stipend from Stiftelsen J. C. Kempes Minnes Stipendiefond, and a travel financing grant from Medicinska Fakulteten Umeå Universitet. The funders had no role in study design, data collection and analysis, decision to publish, or preparation of the article.
Disclosure of potential conflict of interest: M. Bianchi and J. Reichenbach have received research support from the CGD Research Trust. M. J. Niemiec has received research support from the Stiftelsen J. C. Kempes Minnes. U. Siler has received research support from the Swiss National Science Foundation. C. F. Urban has received research support from Molecular Infection Medicine Sweden, Stiftelsen J. C. Kempes Minnes, and Med/Cinska Fakulteten Umea University.
Received for publication October 28, 2010; revised December 26, 2010; accepted for publication January 10, 2011.
Available online March 3, 2011.
Reprint requests: Janine Reichenbach, MD, Division of Immunology/Hematology/BMT, University Children's Hospital Zurich, Steinwiesstrasse 75, 8032 Zurich, Switzerland. E-mail: janine.reichenbach@kispi.uzh.ch. Constantin F. Urban, PhD, Antifungal Immunity Group, Molecular Biology Department, Laboratory for Molecular Infection Medicine Sweden, Umeå University, 90187 Umeå, Sweden. E-mail: constantin. urban@molbiol.umu.se.
0091-6749/$36.00
© 2011 American Academy of Allergy, Asthma & Immunology
doi:10.1016/j.jaci.2011.01.021

1243

Abbreviations used
CGD: Chronic granulomatous disease
DAPI: 4'-6-diamidino-2-phenylindole, dihydrochloride
GT: Gene therapy
LTR: Long terminal repeat
NADPH: Nicotinamide adenine dinucleotide phosphate
NE: Neutrophil elastase
NET: Neutrophil extracellular trap
phox: Phagocyte oxidase
PMA: Phorbol 12-myristate 13-acetate
ROS: Reactive oxygen species

antifungal agent inhibiting *Aspergillus* growth within NETs is the cytoplasmic calprotectin protein complex (S100A8/A9 heteroduplex). This is in good agreement with our previous report showing that calprotectin is the major NET component against *Candida albicans*.[4]

METHODS
We treated a boy 8 years 7 months old with X-linked gp91^{phox} (phox, phagocyte oxidase)-deficient CGD and multifocal therapy–refractory *A nidulans* lung infection with a monocistronic long terminal repeat–driven γ-retroviral SF71gp91^{phox} vector (see reference[11] for details; additional information in this article's Methods section on the Online Repository at www.jacionline.org) under a protocol approved by the ethics review board of the University Children's Hospital Zurich and the Swiss Expert Committee for Bio-Safety after written informed consent of the parents.

NET induction and quantification
Neutrophil extracellular trap formation was quantified as described[11] after stimulation of 5×10^4 neutrophils (control, CGD unsorted, CGD gp91^{phox}-negative [gp91^{phox-}], and gp91^{phox}-positive [gp91^{phox+}] sorted neutrophils) for 4 hours with 40 nmol/L phorbol 12-myristate 13-acetate (PMA, Sigma-Aldrich, Taufkirchen, Germany) and staining the NET-DNA with 1 μmol/L Sytox green (Invitrogen, Molecular Probes, Leiden, The Netherlands) in a black 96-well plate. Nonstimulated neutrophils were used as negative control. The plates were read in a fluorescence microplate reader (Victor 3, PerkinElmer, Waltham, Mass) with a filter setting of 485/535 nm (excitation/emission).

A nidulans growth inhibition
Antifungal activity of purified calprotectin was assessed by incubating *A nidulans* conidia or hyphae with 0.5 to 128 μg/mL recombinant S100A8 and S100A9 (ProtEra, Florence, Italy) ± 1 μmol/L ZnSO$_4$ (Sigma-Aldrich) or 15 μg/mL rabbit polyclonal α-S100A9 antibody (H00006280-D01P, Abnova, Taipei, Taiwan). Antifungal activity of NETs was determined by incubating *A nidulans* conidia or hyphae (strain isolated as described[15] from bronchoalveolar lavage fluid of the patient) with human and murine NETs or human NET extracts ± 0 to 2 μmol/L ZnSO$_4$ or 15 μg/mL rabbit polyclonal α-S100A9 antibody, respectively (additional information in the Methods section on the Online Repository).

NET induction with *A nidulans*
In a 24-well plate, 2×10^5 control, CGD gp91^{phox-} and gp91^{phox+} neutrophils were incubated with *A nidulans* hyphae obtained from culture of 6×10^4 *A nidulans* conidia on polylysine-coated cover slips for 5 hours at 37°C, then fixed in 4% paraformaldehyde for bright field microscopy and immunostaining/confocal microscopy, or 2.5% glutaraldehyde for scanning electron microscopy. For confocal microscopy (SP5; Leica, Wetzlar, Germany), specimens were blocked with 5% donkey serum (Jackson ImmunoResearch, Suffolk, United Kingdom), 1% BSA, 3% cold water fish gelatin (Sigma-Aldrich), and 0.25% Tween 20 in PBS, incubated with primary antibodies directed against S100A8/A9 (BM4029 and BM4027; Acris, Herford, Germany), common *Aspergillus* spp soluble proteins (NB100-65026, Novus Biologicals, Cambridge, United Kingdom), and species-specific secondary antibodies coupled to Alexa Fluor 488 and 568 (Invitrogen, Molecular Probes). DNA was stained with 4'-6-diamidino-2-phenylindole, dihydrochloride (DAPI; Sigma-Aldrich). For scanning electron microscopy (CM208; Philips, Eindhoven, The Netherlands), specimens were postfixed with 2% osmium tetroxide, dehydrated with graded ethanol series (50% to 100%), critical point-dried, and coated with 10 nm platinum.

Release of calprotectin
In a 96-well plate, 5×10^4 control, CGD gp91^{phox-} and gp91^{phox+} neutrophils were stimulated in duplicate with 40 nmol/L PMA at 37°C and 5% CO$_2$ for 4 hours to form NETs. After 30 minutes NET digestion with 5 U/mL Micrococcal nuclease (Worthington Biochemicals, Lakewood, NJ) + 1 U/mL Deoxyribonuclease-1 (Sigma-Aldrich), samples was centrifuged for 10 minutes at 10,000g to remove debris. Heterodimeric calprotectin (S100A8/A9) concentration was measured by ELISA (Hycult, Uden, The Netherlands; detection limit 1.6 ng/mL) with 1/10, 1/40, and 1/80 dilutions. As control, total cytoplasmic calprotectin was quantified after neutrophil lysis.

RESULTS
NET quantification in CGD neutrophils after GT
We aimed to identify factors that limit *A nidulans* growth in NADPH oxidase–dependent NETs. Therefore, we first analyzed whether NET formation was restored 2.6 years after GT in gp91^{phox+} (transduced) compared with gp91^{phox-} (nontransduced) neutrophils from the reported patient with CGD.[11] Staining with α-gp91^{phox} antibody showed 22.8% gp91^{phox+} neutrophils. Sorting of gp91^{phox-} and gp91^{phox+} neutrophils by autoMACS (Miltenyi Biotec, Bergisch Gladbach, Germany) resulted in purity of 98.3% and 92.5% (Fig 1, *A*). Quantification of NET formation in CGD neutrophils after GT correlated well with the percentage of gp91^{phox+} cells, 20.5% of unsorted, 3.2% of gp91^{phox-}, and 87.6% of gp91^{phox+} neutrophils releasing NETs (Fig 1, *B*). The O$_2^-$ production of individual gp91^{phox+} neutrophils was 27.8% (compared with control), measured by reduction of cytochrome c (data not shown),[16] indicating this threshold as enough for substantial, albeit not 100%, NET formation.

Calprotectin inhibits *A nidulans* growth efficiently
We previously identified the cytoplasmic protein complex calprotectin as major antifungal effector in NETs preventing growth of *C albicans*[4] and therefore reasoned that calprotectin might also be responsible for NET-mediated growth inhibition of *A nidulans*. There is currently no information available whether purified calprotectin is able to inhibit *Aspergillus* spp. To evaluate the inhibitory activity of each calprotectin subunit on *A nidulans* growth, we incubated conidia and hyphae with purified subunits S100A8 and S100A9.[17] Addition of S100A8 did not inhibit fungal growth, whereas S100A9 had a small effect on conidia germination but not on hyphae growth (Fig 2, *A* and *B*; see this article's Fig E1 in the Online Repository at www.jacionline.org). Only both subunits together strongly inhibited *A nidulans* growth, indicating that the heteroduplex is required for full inhibition of *Aspergillus* spp, consistent with a report on its anticandidal activity.[18] In all cases, addition of 1 μmol/L zinc (Zn^{2+}) restored fungal growth, confirming the putative role of Zn^{2+} chelation by calprotectin in suppressing fungal growth.[18-20] In addition, we preincubated combined S100A9 and S100A8 with a polyclonal α-S100A9 antibody. *A nidulans* growth of conidia and hyphae

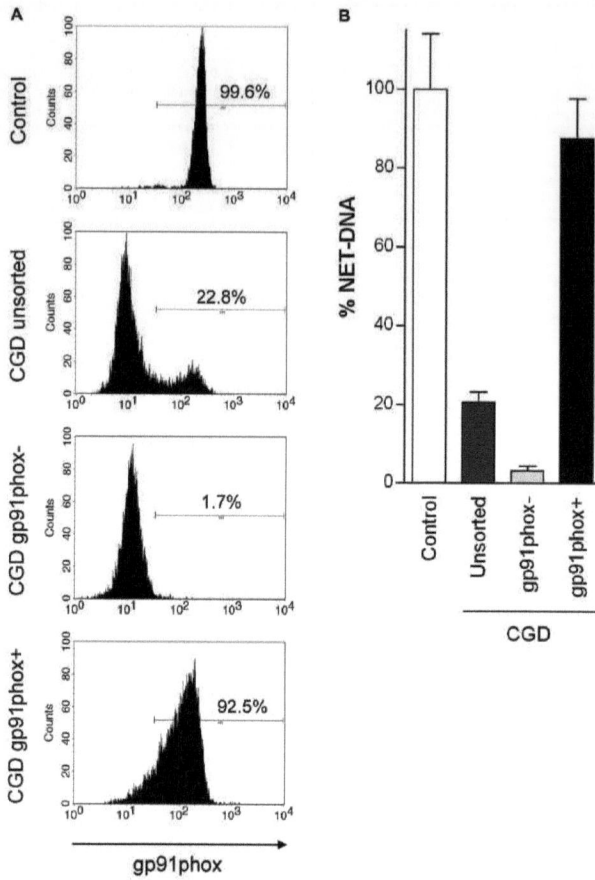

FIG 1. NET quantification in CGD gp91^{phox+} and gp91^{phox-} sorted neutrophils. **A,** Neutrophils were sorted by magnetic beads and autoMACS. **B,** NET formation was quantified after PMA activation of neutrophils by staining of released NET-DNA with Sytox green. Data are means ± SDs of representative triplicate experiments. The difference between control and gp91^{phox+} neutrophils was nonsignificant ($P > .05$).

was restored in the presence of specific antibody (Fig 2, C and D; see this article's Fig E2 in the Online Repository at www.jacionline.org), confirming the requirement for interaction of both calprotectin subunits for antifungal activity. Unspecific control antibody did not have any effect. Here we used an excess amount of 5 μg/mL purified S100A8 and S100A9, but similar inhibitory effects were obtained with as low as 0.5 μg/mL, a calprotectin amount found in NETs from about 4 million neutrophils.[4]

Antifungal activity of CGD gp91^{phox+} cells is calprotectin-dependent

To study whether restored antifungal activity of CGD gp91^{phox+} neutrophils after GT was attributable to calprotectin, we PMA-stimulated gp91^{phox+} and gp91^{phox-} neutrophils to induce NETs, and coincubated with *A nidulans* conidia and hyphae ± Zn^{2+}. Only gp91^{phox+} neutrophils showed strong antifungal activity, equivalent to control (Fig 3, A and B; see this article's

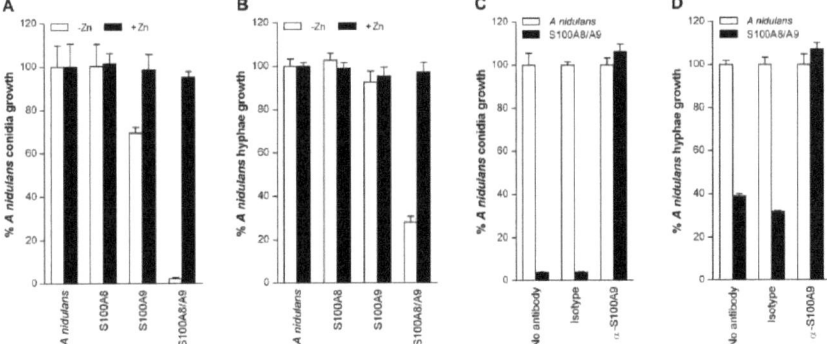

FIG 2. Inhibition of *A nidulans* growth by purified calprotectin. **A and B,** Conidia and hyphae incubated with 5 μg/mL S100A8 and/or S100A9 ± 1 μmol/L Zn^{2+}. **C and D,** Conidia and hyphae incubated with 5 μg/mL S100A8 and S100A9 ± 15 μg/mL polyclonal α-S100A9 antibody or unspecific control antibody. Data are means ± SDs of representative triplicate experiments.

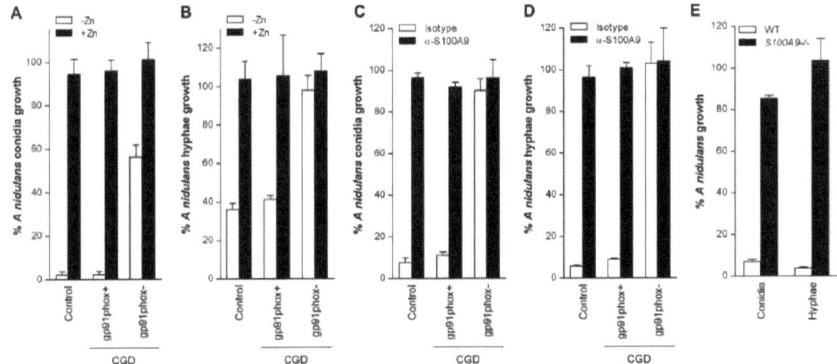

FIG 3. Calprotectin-dependent antifungal activity of CGD gp91[phox+] neutrophils. *A nidulans* conidia or hyphae incubated with NETs **(A and B)** or NET supernatant **(C and D)** from control and CGD-sorted neutrophils ± 1 μmol/L Zn^{2+} or 15 μg/mL α-S100A9, respectively. **E,** Wild-type *(WT)* and *S100A9−/−* mouse NETs infected with *A nidulans* conidia or hyphae. Data are means ± SDs of representative triplicate experiments.

Fig E3 in the Online Repository at www.jacionline.org). gp91[phox−] neutrophils did not form NETs (Fig 1, *B*), yet partially inhibited conidia germination, probably by phagocytosis,[21] but were inefficient against hyphae. Full restoration of *A nidulans* growth by addition of Zn^{2+} supports the assumption that NET antifungal activity was calprotectin-dependent (at calprotectin concentrations of 450-500 ng/mL, measured by ELISA).

Restoration of *A nidulans* conidia and hyphae growth by Zn^{2+} addition was dose-dependent (see this article's Fig E4 in the Online Repository at www.jacionline.org). Addition of Mn^{2+}, Fe^{2+}, and Cu^{2+} (0-2.5 μmol/L), however, did not restore fungal growth (not shown), indicating that Zn^{2+} binding by calprotectin has a major NET anti-*Aspergillus* activity. Because addition of metal ions can have pleiotropic effects, we applied an α-S100A9 antibody to block calprotectin-mediated inhibition in concentrated NET extracts, subsequently incubated with *A nidulans* conidia or hyphae. NET extracts from gp91[phox+] neutrophils in the presence of unspecific control antibody showed strong antifungal activity, similar to control (Fig 3, *C* and *D;* see this article's Fig E5 in the Online Repository at www.jacionline.org). Supernatants from non–NET-forming gp91[phox−] neutrophils in the presence of unspecific control antibody did not have antifungal activity, even against conidia germination, supporting the hypothesis that their low antifungal activity observed against conidia (Fig 3, A) was a result of phagocytosis.

We then tested whether NET-mediated A nidulans inhibition was calprotectin-dependent comparing wild-type and calprotectin-deficient mouse neutrophils. NETs were PMA-induced in mature neutrophils from both mouse strains and incubated with A nidulans conidia and hyphae. Indeed, NETs from calprotectin-deficient neutrophils did not prevent A nidulans growth, whereas NETs from wild-type neutrophils did (Fig 3, E), although the amount of NET formation was similar in calprotectin-deficient and wild-type neutrophils.[4] Complete restoration of fungal growth in the presence of α-S100A9 antibody in human gp91^{phox+} and control NET extracts and the inability of calprotectin-deficient murine NETs to block fungal growth confirmed the essential antifungal role of NET-associated calprotectin against A nidulans.

Antifungal effect of calprotectin

To determine whether calprotectin-mediated inhibition of A nidulans growth within NETs was reversible or irreversible, we incubated conidia and hyphae on NETs for 4 days to 4 weeks at neutral pH, then added Zn^{2+} to restore fungal growth. NETs strongly inhibited A nidulans conidia germination and hyphae growth even after 4 weeks (Fig 4, A-P), indicating strong stability of calprotectin/Zn^{2+} complexes, agreeing well with their reported stability against proteases.[22] Surprisingly, addition of Zn^{2+} restored fungal growth even after 4 weeks, demonstrating that calprotectin has fungistatic but not fungicidal activity on A nidulans at low concentrations (450-500 ng/mL, measured by ELISA).

We next tested whether antifungal activity was still reversible at higher calprotectin concentrations that might occur in vivo. We incubated conidia and hyphae for 4 days with increasing concentrations of purified calprotectin S100A8 and S100A9, then added Zn^{2+} to restore fungal growth. Conidia germination was inhibited at any concentration (Fig 4, Q, white bars), whereas hyphae were slightly more resistant, showing a dose-dependent increase of growth arrest between 0.5 and 4 μg/mL S100A8/A9 (Fig 4, R, white bars). Fungal growth was not restored by Zn^{2+} when the S100A8/A9 concentration was higher than 16 μg/mL (Fig 4, Q-R, black bars). Light microscopy showed intact hyphae at low S100A8/A9 concentrations (Fig 4, S), whereas in the presence of >16 μg/mL S100A8/A9, hyphae had a translucent cell wall and a fragmented internal structure (Fig 4, T). Thus, we demonstrate a previously unknown microbicidal effect against A nidulans at high calprotectin concentrations.

A nidulans–induced NET formation is strictly dependent on NADPH oxidase

We investigated whether the induction of NETs by A nidulans is dependent on functional NADPH oxidase and is restored after CGD GT. CGD gp91^{phox+} and gp91^{phox-} neutrophils were incubated with A nidulans hyphae and observed for NET formation by light microscopy, scanning electron microscopy, and immunostaining with confocal microscopy. Hyphae induced NETs in control and gp91^{phox+} neutrophils but, as expected, less strongly compared with PMA stimulation. We observed that NET formation mainly occurred when neutrophils were in direct contact with hyphae (Fig 5, A and B, black arrows). This might suggest that close pathogen-neutrophil interaction promotes induction of NETs. The molecular details behind this trigger mechanism will be subject to further investigation. In contrast, gp91^{phox-} neutrophils did not form NETs when in contact with hyphae (Fig 5, C, white arrowheads), underlining the role of functional NADPH oxidase in NET formation and antifungal defense. Analysis by scanning electron microscopy showed that A nidulans hyphae were entangled by NETs from control and gp91^{phox+} neutrophils (Fig 5, D, G, and E, H), whereas gp91^{phox-} neutrophils surrounded and wrapped around hyphae but did not make NETs (Fig 5, F, I).

Calprotectin is NET-associated after A nidulans incubation

The association of calprotectin and A nidulans–induced NETs was studied by immunofluorescence and confocal microscopy, staining DNA with DAPI and calprotectin and hyphae with primary antibodies. Colocalization of DNA (blue) and calprotectin (red) demonstrates that calprotectin was released and associated with NETs when control (Fig 5, J, M, P) and gp91^{phox+} (Fig 5, K, N, Q) neutrophils were incubated with A nidulans hyphae (green). Moreover, we demonstrate that gp91^{phox-} neutrophils showed neither NET formation nor extracellular calprotectin on A nidulans incubation (Fig 5, L, O, R), although they were able to recognize A nidulans, as judged by migration toward and wrapping around hyphae (Fig 5, R). DNA was confined to the nucleus (Fig 5, L) and calprotectin to the cytoplasmic compartment of these cells (Fig 5, O).

Notably, the quantification of calprotectin by ELISA confirmed our microscopic findings. Control, gp91^{phox+} and gp91^{phox-} neutrophils were activated with PMA for 4 hours, and calprotectin release was determined by ELISA after NET formation and subsequent digestion of NETs. Control and gp91^{phox+} neutrophils released approximately 4 times the calprotectin concentration compared with gp91^{phox-} neutrophils (Fig 6), supporting the previous results. Differences in total cytoplasmic calprotectin concentration among control, gp91^{phox+}, and gp91^{phox-} neutrophils were not statistically significant (data not shown). Thus, we conclude that NADPH oxidase is required not only for the release of NETs but also for the release of calprotectin, the crucial component for the inhibition of A nidulans hyphae.

DISCUSSION

Recent work suggests NET contribution to elimination of fungal infections in healthy subjects because hyphae are too large to be phagocytosed.[23,24] Patients with CGD are unable to make NETs[5,11] and are susceptible to therapy-refractory infection with Aspergillus spp.[10] Especially infections with A nidulans are a major threat to these patients because of higher virulence and frequent resistance to antifungal treatment compared to Aspergillus fumigatus.[7] Here we show that genetic complementation of NADPH oxidase by human CGD GT restored the ability of neutrophils to release NETs and NET-associated calprotectin, which is responsible for growth inhibition of A nidulans by Zn^{2+} sequestration. A threshold of 27.8% of O_2^- production per individual gp91^{phox+} neutrophil and a chimerism of 22.8% of oxidase-positive cells was enough for sufficient NET formation in vitro and clinical clearance of multifocal therapy-refractory A nidulans lung infection in a patient with CGD in vivo.

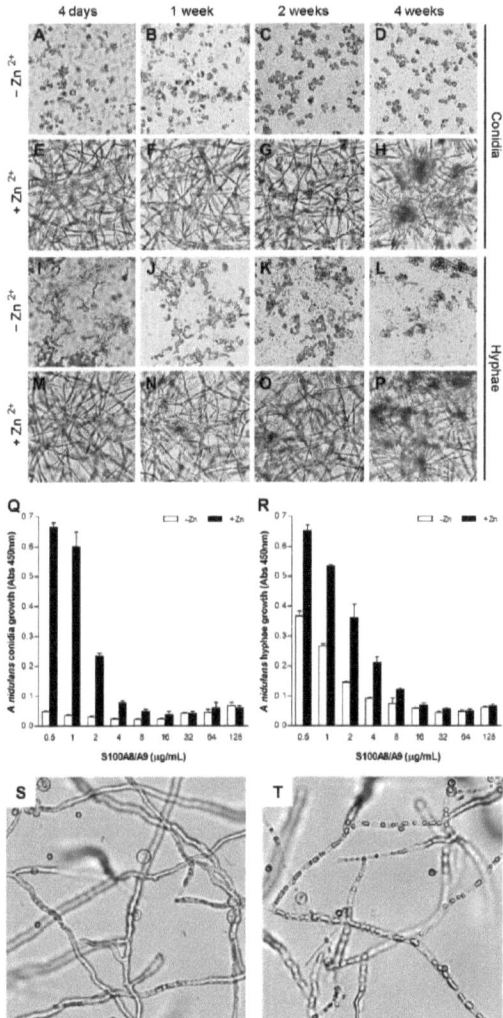

FIG 4. Antifungal effect of calprotectin. A-P, Fungistatic activity: *A nidulans* conidia or hyphae incubated with NETs for 4 days to 4 weeks, after which 1 μmol/L Zn^{2+} was added overnight. Q and R, Fungicidal activity: *A nidulans* conidia or hyphae incubated with 0.5 (S) to 128 (T) μg/mL purified S100A8/A9 for 4 days, after which 50 μmol/L Zn^{2+} was added overnight.

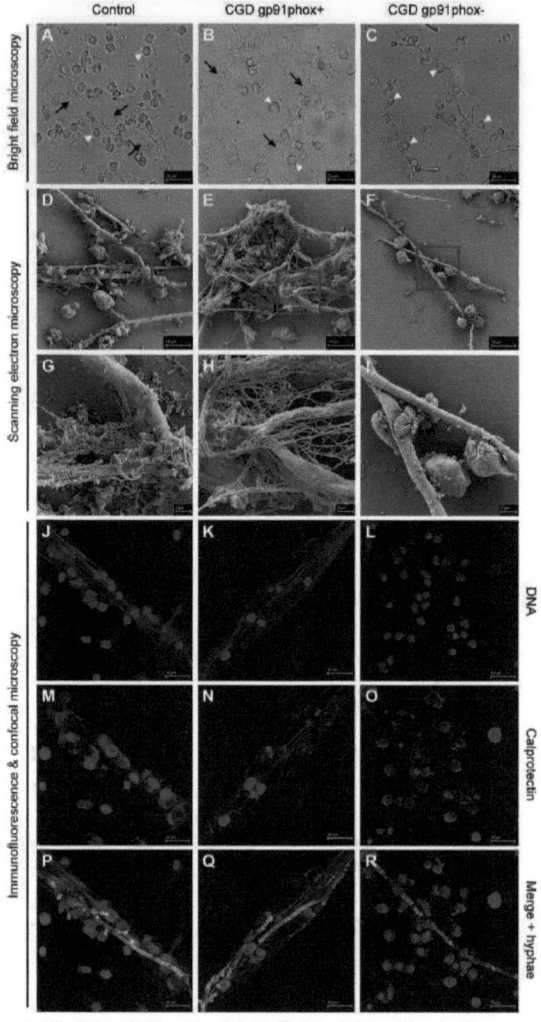

FIG 5. *A nidulans*-induced NETs contain calprotectin. Control and CGD sorted neutrophils were infected with hyphae, and NET formation was determined by bright field microscopy (A-C), scanning electron microscopy (D-I), and immunofluorescence and confocal microscopy (J-R). NET formation was determined by DAPI *(blue)*, calprotectin by staining with antibodies against S100A8/A9 *(red)*, and hyphae with antibodies against *Aspergillus* soluble proteins *(green)*.

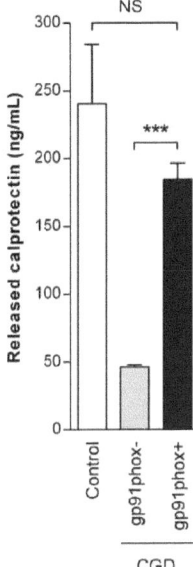

FIG 6. Calprotectin release by gp91phox+ and gp91phox− neutrophils. Calprotectin release was quantified by ELISA after neutrophil activation by PMA inducing NET formation. NETs were digested by MNase/DNase-1 to quantify NET-associated calprotectin. Significance was assessed by the Student t test: NS, not significant ($P > .05$); ***$P < .001$.

The S100 Ca^{2+}-binding protein calprotectin is a heteroduplex of subunits S100A8 and S100A9 that is expressed by granulocytes, monocytes, and early differentiation stages of macrophages.[25] It represents about 40 % of neutrophil cytoplasm,[26] accumulates at high concentrations (1-20 mg/mL) in abscess fluid supernatants,[27] and has been shown to bind Zn^{2+} in vitro.[28] Its growth-inhibitory effect on fungi is most likely mediated by metal chelation and is reversible by micromolar quantities of Zn^{2+}.[18,27,28] We set out to study the role of calprotectin in innate defense against CGD-relevant A nidulans. Growth of A nidulans was strongly inhibited when coincubated with both purified calprotectin subunits, which is in accordance with a report by Sohnle et al[17] and with crystal structure[29] as well as mass spectrometric analyses[30] suggesting that the Ca^{2+}-dependent formation of S100A8/A9 heterotetramers would lead to enhanced Zn^{2+}-binding capacity, whereas none of the homodimers should display significant affinity for Zn^{2+}.[29]

Our study demonstrates the major role of NET-associated calprotectin in suppressing growth of A nidulans conidia and hyphae and shows the presence of calprotectin in NETs induced by A nidulans hyphae. A fumigatus has been reported recently to trigger NET release,[31,32] resulting in growth inhibition of A fumigatus that could be prevented by Zn^{2+}.[32] However, the mechanism of inhibition was not clearly shown in these reports. Here we show that blocking of calprotectin by specific antibodies and the absence of calprotectin in NETs from calprotectin-deficient mouse neutrophils completely abrogated the inhibition of A nidulans conidia and hyphae growth. This is a more precise analysis because NETs also contain other Zn^{2+}-binding proteins, such as S100A12,[33] which is involved in antiparasite responses.

According to our findings, calprotectin has to be presented extracellularly for antifungal activity against A nidulans. To the best of our knowledge, leakage of a cytoplasmic protein into phagolysosomes has not been described. We previously showed that neutrophil degranulation stimulated by the bacterial peptide formyl-Methionyl-Leucyl-Phenylalanine does not induce secretion of calprotectin, indicating nonvesicular localization.[4] However, we cannot entirely exclude that minor amounts of calprotectin are found in phagolysosomes and contribute to the slight inhibitory effect we reported on CGD neutrophils infected with A nidulans conidia.

We show that the presence of functional NADPH oxidase in human neutrophils is an absolute requirement for A nidulans growth inhibition by NET-associated calprotectin, because only gp91phox+ and not gp91phox− neutrophils made NETs containing calprotectin and inhibited A nidulans growth efficiently. GT of human CGD restored the amount of released calprotectin during NET formation to wild-type levels, and NET-associated calprotectin colocalized with A nidulans hyphae. Analysis of gp91phox− neutrophils enabled us to establish the role of NADPH oxidase for the release of calprotectin during NET formation, which is an advantage over chemical inhibition of NADPH oxidase using diphenylene iodonium chloride. We propose that NET formation is an NADPH oxidase–dependent mechanism by which neutrophils secrete calprotectin into the extracellular compartment on infection with Aspergillus hyphae, resulting in both NET-bound and free calprotectin. NET-associated calprotectin is concentrated with trapped Aspergillus for antifungal function. Unbound calprotectin may in addition have distinct functions, such as chemotaxis or in inflammation.

The fact that gp91phox− neutrophils also partially inhibited conidia germination suggests that NADPH oxidase–independent mechanisms linked to phagocytosis, such as activity of the granule serine proteases NE and cathepsin G[34-36] (although the mechanism has been recently object of controversy[37-39]), might contribute to early defense against Aspergillus. The need of NE for NET formation[14] could explain the susceptibility to infection of NE-deficient mice described by Reeves et al[35] as an indirect action. Other mechanisms such as the tryptophan-catabolizing enzyme indoleamine-2,3-dioxygenase, considered critical for regulating immune responses and suppression of inflammation in mice,[40] do not play a role in human CGD as we previously reported.[41,42]

Fungi have a high Zn^{2+} requirement (10^{-7}-10^{-5} mol/L) for growth[43] with a low minimum inhibitory concentration of calprotectin.[44] Because of the lack of a well developed extracellular Zn^{2+}-scavenging system, A fumigatus is extremely sensitive to Zn^{2+} deprivation.[43] The Zn^{2+} requirement for A nidulans in our study was in the same range as described for other fungi. In contrast with Staphylococcus aureus, which is more sensitive to calprotectin-mediated Mn^{2+} deprivation,[19] we found that A nidulans could resume growth only after Zn^{2+} supplementation, but not after Mn^{2+}, Fe^{2+}, or Cu^{2+} addition. This supports the central role of Zn^{2+} for A nidulans growth.

Interestingly, *A nidulans* hyphae were not killed by NETs at low calprotectin concentrations *in vitro*, although hyphal structures seemed to be severely deranged by NETs. This corresponded to the range of fungistatic calprotectin concentrations described for *Candida*.[44] *A nidulans* growth was restored after addition of Zn^{2+}, even after a month of inhibition by NETs. Because much higher calprotectin concentrations have been observed in human abscesses,[27] *Aspergillus* hyphae might still be killed by NET-associated calprotectin *in vivo*. Killing of *Candida* spp or *Cryptococcus neoformans* was demonstrated at calprotectin concentrations of 3 to 5 μg/mL.[44] We show for the first time that calprotectin concentrations ≥16 μg/mL lead to irreversible growth arrest of *A nidulans* conidia and hyphae, most likely by Zn^{2+} starvation. In contrast with our findings, a previous publication reported only minor antifungal activity of NETs against *A fumigatus*,[31] probably because of a shorter incubation period. Alternatively, complete *in vivo* killing might also be achieved by other antimicrobial peptides[46] or by so far unknown mechanisms: in addition to their antimicrobial properties, S100A8, S100A9, and S100A8/A9 are chemotactic factors for neutrophils and monocytes.[47-49] Notably, for this function, calprotectin needs to be extracellular, and thus *Aspergillus*-induced NET formation supports it as well.

Here we establish that calprotectin released in NETs displays concentration-dependent reversible and irreversible growth inhibitory activity against virulent *A nidulans*. NETs may be involved in disarming *A nidulans* and may prevent further spreading by providing high local concentrations of calprotectin. Definitive *in vivo* proof in human beings is obviously ethically impossible. In conclusion, we show that calprotectin is a critical factor in the innate immune defense of human neutrophils to *Aspergillus* infection and adds to the concept of metal chelation as a strategy for inhibiting microbial growth at the sites of infection. The severe immunodeficient phenotype and the high susceptibility to *Aspergillus* infection of patients with CGD might be linked to absence of NETs, and restoration of NADPH oxidase function and NET formation by GT leads to rapid cure of refractory invasive aspergillosis in X-linked CGD.

We thank the patient and his family for their trust. We are indebted to the medical and nursing staff of the bone marrow transplantation unit of University Children's Hospital Zurich. We thank Reinhard Seger for his support and helpful discussions, and Manuel Grez and Klaus Kühlke for developing and providing the SF71gp91phox vector, respectively. We are grateful to Klaus Marquardt for electron microscopy, to Andres Kaech for help with confocal microscopy, and to Alex Imhof for isolating *A nidulans* conidia. Furthermore, we acknowledge Thomas Vogl and Johannes Roth for supplying S100A9 knockout mice.

Clinical implications: Gene therapy for CGD restores NET formation and calprotectin release by neutrophils, leading to efficient *A nidulans* inhibition. Calprotectin is a critical factor in human innate immune defense against *Aspergillus* infection.

REFERENCES

1. Brinkmann V, Reichard U, Goosmann C, Fauler B, Uhlemann Y, Weiss DS, et al. Neutrophil extracellular traps kill bacteria. Science 2004;303:1532-5.
2. Guimaraes-Costa AB, Nascimento MT, Froment GS, Soares RP, Morgado FN, Conceicao-Silva F, et al. Leishmania amazonensis promastigotes induce and are killed by neutrophil extracellular traps. Proc Natl Acad Sci U S A 2009;106:6748-53.
3. Urban CF, Reichard U, Brinkmann V, Zychlinsky A. Neutrophil extracellular traps capture and kill Candida albicans yeast and hyphal forms. Cell Microbiol 2006;8:668-76.
4. Urban CF, Ermert D, Schmid M, Abu-Abed U, Goosmann C, Nacken W, et al. Neutrophil extracellular traps contain calprotectin, a cytosolic protein complex involved in host defense against Candida albicans. PLoS Pathog 2009;5:e1000639.
5. Fuchs TA, Abed U, Goosmann C, Hurwitz R, Schulze I, Wahn V, et al. Novel cell death program leads to neutrophil extracellular traps. J Cell Biol 2007;176:231-41.
6. Hsu U, Champaiboon C, Guenther BD, Sorenson BS, Khammanivong A, Ross KF, et al. Anti-infective protective properties of S100 calgranulins. Antiinflamm Antiallergy Agents Med Chem 2009;8:290-305.
7. Segal BH, DeCarlo ES, Kwon-Chung KJ, Malech HL, Gallin JI, Holland SM. Aspergillus nidulans infection in chronic granulomatous disease. Medicine (Baltimore) 1998;77:345-54.
8. Seger RA. Modern management of chronic granulomatous disease. Br J Haematol 2008;140:255-66.
9. Winkelstein JA, Marino MC, Johnston RB Jr, Boyle J, Curnutte J, Gallin JI, et al. Chronic granulomatous disease: report on a national registry of 368 patients. Medicine (Baltimore) 2000;79:155-69.
10. Segal BH, Romani LR. Invasive aspergillosis in chronic granulomatous disease. Med Mycol 2009;47(suppl 1):S282-90.
11. Bianchi M, Hakkim A, Brinkmann V, Siler U, Seger RA, Zychlinsky A, et al. Restoration of NET formation by gene therapy in CGD controls aspergillosis. Blood 2009;114:2619-22.
12. Ermert D, Urban CF, Laube B, Goosmann C, Zychlinsky A, Brinkmann V. Mouse neutrophil extracellular traps in microbial infections. J Innate Immun 2009;1:181-93.
13. Metzler KD, Fuchs TA, Nauseef WM, Reumaux D, Roesler J, Schulze I, et al. Myeloperoxidase is required for neutrophil extracellular trap formation: implications for innate immunity. Blood 2010;117:953-9.
14. Papayannopoulos V, Metzler KD, Hakkim A, Zychlinsky A. Neutrophil elastase and myeloperoxidase regulate the formation of neutrophil extracellular traps. J Cell Biol 2010;191:677-91.
15. Bonnett CR, Cornish EJ, Harmsen AG, Burritt JB. Early neutrophil recruitment and aggregation in the murine lung inhibit germination of Aspergillus fumigatus Conidia. Infect Immun 2006;74:6528-39.
16. Mayo LA, Curnutte JT. Kinetic microplate assay for superoxide production by neutrophils and other phagocytic cells. Methods Enzymol 1990;186:567-75.
17. Sohnle PG, Hunter MJ, Hahn B, Chazin WJ. Zinc-reversible antimicrobial activity of recombinant calprotectin (migration inhibitory factor-related proteins 8 and 14). J Infect Dis 2000;182:1272-5.
18. Murthy AK, Lehrer RI, Harwig SS, Miyasaki KT. In vitro candidastatic properties of the human neutrophil calprotectin complex. J Immunol 1993;151:6291-301.
19. Corbin BD, Seeley EH, Raab A, Feldmann J, Miller MR, Torres VJ, et al. Metal chelation and inhibition of bacterial growth in tissue abscesses. Science 2008;319:962-5.
20. Sohnle PG, Hahn BL, Santhanagopalan V. Inhibition of Candida albicans growth by calprotectin in the absence of direct contact with the organisms. J Infect Dis 1996;174:1369-72.
21. Behnsen J, Narang P, Hasenberg M, Gunzer F, Bilitewski U, Klippel N, et al. Environmental dimensionality controls the interaction of phagocytes with the pathogenic fungi Aspergillus fumigatus and Candida albicans. PLoS Pathog 2007;3:e13.
22. Nacken W, Kerkhoff C. The hetero-oligomeric complex of the S100A8/S100A9 protein is extremely protease resistant. FEBS Lett 2007;581:5127-30.
23. Latge JP. Aspergillus fumigatus and aspergillosis. Clin Microbiol Rev 1999;12:310-50.
24. Mizgerd JP. Acute lower respiratory tract infection. N Engl J Med 2008;358:716-27.
25. Ehrchen JM, Sunderkotter C, Foell D, Vogl T, Roth J. The endogenous Toll-like receptor 4 agonist S100A8/S100A9 (calprotectin) as innate amplifier of infection, autoimmunity, and cancer. J Leukoc Biol 2009;86:557-66.
26. Edgeworth J, Gorman M, Bennett R, Freemont P, Hogg N. Identification of p8,14 as a highly abundant heterodimeric calcium binding protein complex of myeloid cells. J Biol Chem 1991;266:7706-13.
27. Clohessy PA, Golden BE. Calprotectin-mediated zinc chelation as a biostatic mechanism in host defence. Scand J Immunol 1995;42:551-6.
28. Sohnle PG, Collins-Lech C, Wiessner JH. The zinc-reversible antimicrobial activity of neutrophil lysates and abscess fluid supernatants. J Infect Dis 1991;164:137-42.
29. Korndorfer IP, Brueckner F, Skerra A. The crystal structure of the human (S100A8/S100A9)2 heterotetramer, calprotectin, illustrates how conformational changes of interacting alpha-helices can determine specific association of two EF-hand proteins. J Mol Biol 2007;370:887-98.
30. Vogl T, Leukert N, Barczyk K, Strupat K, Roth J. Biophysical characterization of S100A8 and S100A9 in the absence and presence of bivalent cations. Biochim Biophys Acta 2006;1763:1298-306.
31. Bruns S, Kniemeyer O, Hasenberg M, Aimanianda V, Nietzsche S, Thywissen A, et al. Production of extracellular traps against Aspergillus fumigatus in vitro and in

infected lung tissue is dependent on invading neutrophils and influenced by hydrophobin RodA. PLoS Pathog 2010;6:e1000873.
32. McCormick A, Heesemann L, Wagener J, Marcos V, Hartl D, Loeffler J, et al. NETs formed by human neutrophils inhibit growth of the pathogenic mold Aspergillus fumigatus. Microbes Infect 2010;12:928-36.
33. Moroz OV, Blagova EV, Wilkinson AJ, Wilson KS, Bronstein IB. The crystal structures of human S100A12 in apo form and in complex with zinc: new insights into S100A12 oligomerisation. J Mol Biol 2009;391:536-51.
34. Decleva E, Menegazzi R, Busetto S, Patriarca P, Dri P. Common methodology is inadequate for studies on the microbicidal activity of neutrophils. J Leukoc Biol 2006;79:87-94.
35. Reeves EP, Lu H, Jacobs HL, Messina CG, Bolsover S, Gabella G, et al. Killing activity of neutrophils is mediated through activation of proteases by K^+ flux. Nature 2002;416:291-7.
36. Tkalcevic J, Novelli M, Phylactides M, Iredale JP, Segal AW, Roes J. Impaired immunity and enhanced resistance to endotoxin in the absence of neutrophil elastase and cathepsin G. Immunity 2000;12:201-10.
37. Ahluwalia J, Tinker A, Clapp LH, Duchen MR, Abramov AY, Pope S, et al. The large-conductance $Ca2+$-activated $K+$ channel is essential for innate immunity. Nature 2004;427:853-8.
38. Essin K, Salanova B, Kettritz R, Sausbier M, Luft FC, Kraus D, et al. Large-conductance calcium-activated potassium channel activity is absent in human and mouse neutrophils and is not required for innate immunity. Am J Physiol Cell Physiol 2007;293:C45-54.
39. Femling JK, Cherny VV, Morgan D, Rada B, Davis AP, Czirjak G, et al. The antibacterial activity of human neutrophils and eosinophils requires proton channels but not BK channels. J Gen Physiol 2006;127:659-72.
40. Romani L, Fallarino F, De Luca A, Montagnoli C, D'Angelo C, Zelante T, et al. Defective tryptophan catabolism underlies inflammation in mouse chronic granulomatous disease. Nature 2008;451:211-5.
41. Hakkim A, Hurwitz R, Bianchi M, Brinkmann V, Siler U, Seger R, et al. Protecting against Aspergillus infection in CGD. Blood 2009;114:3498.
42. Jurgens B, Fuchs D, Reichenbach J, Heitger A. Intact indoleamine 2,3-dioxygenase activity in human chronic granulomatous disease. Clin Immunol 2010;137: 1-4.
43. Sugarman B. Zinc and infection. Rev Infect Dis 1983;5:137-47.
44. Steinbakk M, Naess-Andresen CF, Lingaas E, Dale I, Brandtzaeg P, Fagerhol MK. Antimicrobial actions of calcium binding leucocyte L1 protein, calprotectin. Lancet 1990;336:763-5.
45. Lulloff SJ, Hahn BL, Sohnle PG. Fungal susceptibility to zinc deprivation. J Lab Clin Med 2004;144:208-14.
46. Levitz SM, Selsted ME, Ganz T, Lehrer RI, Diamond RD. In vitro killing of spores and hyphae of Aspergillus fumigatus and Rhizopus oryzae by rabbit neutrophil cationic peptides and bronchoalveolar macrophages. J Infect Dis 1986; 154:483-9.
47. Ryckman C, Vandal K, Rouleau P, Talbot M, Tessier PA. Proinflammatory activities of S100: proteins S100A8, S100A9, and S100A8/A9 induce neutrophil chemotaxis and adhesion. J Immunol 2003;170:3233-42.
48. Vandal K, Rouleau P, Boivin A, Ryckman C, Talbot M, Tessier PA. Blockade of S100A8 and S100A9 suppresses neutrophil migration in response to lipopolysaccharide. J Immunol 2003;171:2602-9.
49. Raquil MA, Anceriz N, Rouleau P, Tessier PA. Blockade of antimicrobial proteins S100A8 and S100A9 inhibits phagocyte migration to the alveoli in streptococcal pneumonia. J Immunol 2008;180:3366-74.

METHODS
Patient description and GT
We treated a boy 8 years 7 months old with X-linked gp91[phox]-deficient CGD and multifocal therapy–refractory *A nidulans* lung infection with a monocistronic long terminal repeat-driven γ-retroviral SF71gp91[phox] vector (see reference[E1] for details) under a protocol approved by the ethics review board of the University Children's Hospital Zurich and the Swiss Expert Committee for Bio-Safety after written informed consent of the parents. Collection of CD34[+] cells, transduction, conditioning with low-dose intravenous busulfan (8.8 mg/kg), and clinical follow-up were performed as described.[E2] The *A nidulans* lung infection was completely cleared 6 weeks after GT. Therapy/prophylaxis with oral voriconazole was continued throughout and has not yet been tapered 2.6 years after GT. To avoid interactions, voriconazole was stopped 4 days before *Aspergillus* experiments and continued thereafter. At the time of analysis, the patient was free of any infection, and 22.8% of neutrophils expressed gp91[phox], with an O_2^- production of 27.8% per gp91[phox]-expressing cell (relative to healthy controls).[E3] Clinical and molecular follow-up of GT in this patient will be described elsewhere.

A nidulans strain
The *A nidulans* strain used was isolated from bronchoalveolar lavage fluid of the patient; conidia were grown and collected as described.[E4] For all experiments, conidia and hyphae were grown in serum-free RPMI medium (phenol red–free) supplemented with 10 mmol/L HEPES. If not stated otherwise, fungal growth was quantified in all assays with the tetrazolium dye 2,3-bis (2-methoxy-4-nitro-5-sulfophenyl)-2H-tetrazolium-5-carboxanilide (XTT, Invitrogen, Molecular Probes, Leiden, The Netherlands) as described.[E5] This method, unlike colony-forming unit enumeration, does not require dispersion of the hyphae.

Isolation of human and murine neutrophils
Human neutrophils were isolated from peripheral blood of healthy donors or the patient with CGD by using dextran-Ficoll (GE Healthcare, Munich, Germany)[E6] for all NET assays with *A nidulans* and for autoMACS sorting, whereas Percoll (GE Healthcare) density gradient separation[E7] was used for NET quantification assays. For all experiments, neutrophils were resuspended in serum-free RPMI medium (phenol red–free) supplemented with 10 mmol/L HEPES and used within 1 hour after isolation.

Murine neutrophils were isolated from *S100A9*[−/−] mice backcrossed 10 times into C57 BL/6. These mice are deficient in both calprotectin subunits S100A8 and S100A9 protein.[E8] Mice were bred in our animal facility according to regulations of the Jordbruksverket Sweden (Dnr A29-69). Mature murine neutrophils were isolated from bone marrow as previously described.[E7] Briefly, bone marrow cells from tibia and femur were singularized by using a 70-μm cell strainer and separated by centrifugation for 30 minutes at 1500g on a discontinuous Percoll gradient with 52% (vol/vol), 69% (vol/vol), and 78% (vol/vol). Neutrophils harvested from the distinct layer between 69% and 78% were resuspended in HBSS without Ca^{2+} and Mg^{2+}.

Sorting of CGD gp91[phox+] and gp91[phox−] neutrophils
Control and patient neutrophils were first stained for 30 minutes at room temperature with 10 μg/mL mouse α-gp91[phox]–fluorescein isothiocyanate antibody (clone 7D5; MBL, Woburn, Md), then 30 minutes at 4°C with anti–fluorescein isothiocyanate MicroBeads (Miltenyi Biotec, Bergisch Gladbach, Germany) for magnetic separation by autoMACS (Miltenyi Biotec), according to the manufacturer's instructions. Purity of the resulting gp91[phox−] and gp91[phox+] populations was directly measured by the fluorescent-activated cell sorter.

A nidulans growth inhibition by purified calprotectin
In a 96-well plate, 10^4 *A nidulans* conidia or hyphae obtained from culture of 10^4 conidia were incubated 24 hours at 37°C with 5 μg/mL recombinant S100A8 and/or 5 μg/mL recombinant S100A9 (ProtEra, Florence, Italy) ± 1 μmol/L $ZnSO_4$ (Sigma-Aldrich, Taufkirchen, Germany) or 15 μg/mL rabbit polyclonal α-S100A9 antibody (H00006280-D01P; Abnova, Taipei, Taiwan) versus unspecific control antibody (Sigma-Aldrich).[E10] This α-S100A9 antibody recognizes the whole S100A9 subunit alone or the S100A9 subunit complexed with S100A8. Fungal growth is expressed as percentage of control values (*A nidulans* conidia or hyphae incubated in media without neutrophils ± $ZnSO_4$, S100A8/A9, or antibodies).

A nidulans growth inhibition by NETs
In a 96-well plate, 10^5 control, CGD gp91[phox−], and gp91[phox+] sorted neutrophils were activated with 40 nmol/L PMA (Sigma-Aldrich) at 37°C and 5% CO_2 for 4 hours until NET formation was complete. Zero to 2 μmol/L $ZnSO_4$ (Sigma-Aldrich) and *A nidulans* conidia or hyphae (hyphae previously grown overnight in a 24-well plate, collected by pipetting and scraping the wells with a cell scraper) with a multiplicity of infection *A nidulans*/neutrophils of 0.1 were then added, in a final volume of 160 μL. Afterward, the plates were centrifuged for 5 minutes at 400g and incubated for 24 hours at 37°C, allowing conidia germination and hyphal outgrowth. Fungal growth is expressed as percentage of control values (*A nidulans* conidia or hyphae incubated in media ± $ZnSO_4$).

A nidulans growth inhibition by NET extracts
Neutrophil extracellular trap formation was induced in 8×10^5 control, CGD gp91[phox−], and gp91[phox+] neutrophils by activation with 40 nmol/L PMA at 37°C and 5% CO_2 for 5 hours in a 24-well plate (500 μL/well; 4 wells each; 3.2×10^6 cells total). For maximal DNA fragmentation, NETs were then digested for 30 minutes with 5 U/mL MNase (Worthington Biochemical, Lakewood, NJ) + 1 U/mL DNase-1 (Sigma-Aldrich) and pooled supernatant concentrated ~10-fold by centrifugation on filter columns with a 3-kd cutoff (Amicon; Millipore, Zug, Switzerland), following the manufacturer's instructions. In a 96-well plate, 10^4 *A nidulans* conidia or hyphae obtained from culture of 6×10^4 *A nidulans* conidia were then incubated for 24 hours at 37°C with the concentrated NET extracts (diluted 1/3) with or without 15 μg/mL α-S100A9 or unspecific control antibody in 100 μL final volume. Fungal growth is expressed as percentage of control values (*A nidulans* conidia or hyphae incubated in media ± antibodies).

A nidulans growth inhibition by murine *S100A9*[−/−] NETs
In a 24-well plate, 5×10^5 mouse neutrophils in 500 μL RPMI with 1% (vol/vol) mouse serum were stimulated with 100 nmol/L PMA for 20 hours at 37°C with 5% CO_2 to induce NET formation. The supernatant was discarded; NETs were washed once with RPMI and incubated with 500 μL RPMI containing *A nidulans* conidia or hyphae at multiplicities of infection 0.1 and 0.01, respectively. Afterward, plates were centrifuged for 5 minutes at 300g and incubated for 20 hours at 37°C. Fungal growth is expressed as percentage of control values (*A nidulans* conidia or hyphae incubated in media).

Fungistatic and fungicidal effect of calprotectin
In a 96-well plate, 10^4 *A nidulans* conidia or hyphae were incubated on previously formed NETs (from 10^5 neutrophils activated with 40 nmol/L PMA at 37°C and 5% CO_2 for 4 hours) or with 0.5 to 128 μg/mL purified S100A8/A9. After 4 days to 4 weeks, RPMI or $ZnSO_4$ (final concentration, 1 μmol/L for NETs and 50 μmol/L for purified S100A8/A9) were added to the wells and incubated overnight. Growth of *A nidulans* conidia or hyphae was assessed by using an inverted light microscope (Leica DM IL HC Fluo, Wetzlar, Germany) coupled to a CCD camera (Leica DFC 480 R2).

Statistical analysis
A 2-tailed Student *t* test was used for analysis of 2 groups. Differences were considered statistically significant when $P < .05$. All statistical tests were performed by using GraphPad Prism.

REFERENCES

E1. Bianchi M, Hakkim A, Brinkmann V, Siler U, Seger RA, Zychlinsky A, et al. Restoration of NET formation by gene therapy in CGD controls aspergillosis. Blood 2009;114:2619-22.
E2. Ott MG, Schmidt M, Schwarzwaelder K, Stein S, Siler U, Koehl U, et al. Correction of X-linked chronic granulomatous disease by gene therapy, augmented by insertional activation of MDS1-EVI1, PRDM16 or SETBP1. Nat Med 2006;12:401-9.
E3. Mayo LA, Curnutte JT. Kinetic microplate assay for superoxide production by neutrophils and other phagocytic cells. Methods Enzymol 1990;186:567-75.
E4. Bonnett CR, Cornish EJ, Harmsen AG, Burritt JB. Early neutrophil recruitment and aggregation in the murine lung inhibit germination of Aspergillus fumigatus conidia. Infect Immun 2006;74:6528-39.
E5. Meshulam T, Levitz SM, Christin L, Diamond RD. A simplified new assay for assessment of fungal cell damage with the tetrazolium dye, (2,3)-bis-(2-methoxy-4-nitro-5-sulphenyl)-(2H)-tetrazolium-5-carboxanilide (XTT). J Infect Dis 1995;172:1153-6.
E6. Weiss J, Kao L, Victor M, Elsbach P. Oxygen-independent intracellular and oxygen-dependent extracellular killing of Escherichia coli S15 by human polymorphonuclear leukocytes. J Clin Invest 1985;76:206-12.
E7. Aga E, Katschinski DM, van Zandbergen G, Laufs H, Hansen B, Muller K, et al. Inhibition of the spontaneous apoptosis of neutrophil granulocytes by the intracellular parasite Leishmania major. J Immunol 2002;169:898-905.
E8. Manitz MP, Horst B, Seeliger S, Strey A, Skryabin BV, Gunzer M, et al. Loss of S100A9 (MRP14) results in reduced interleukin-8-induced CD11b surface expression, a polarized microfilament system, and diminished responsiveness to chemoattractants in vitro. Mol Cell Biol 2003;23:1034-43.
E9. Ermert D, Urban CF, Laube B, Goosmann C, Zychlinsky A, Brinkmann V. Mouse neutrophil extracellular traps in microbial infections. J Innate Immun 2009;1:181-93.
E10. Sohnle PG, Hunter MJ, Hahn B, Chazin WJ. Zinc-reversible antimicrobial activity of recombinant calprotectin (migration inhibitory factor-related proteins 8 and 14). J Infect Dis 2000;182:1272-5.

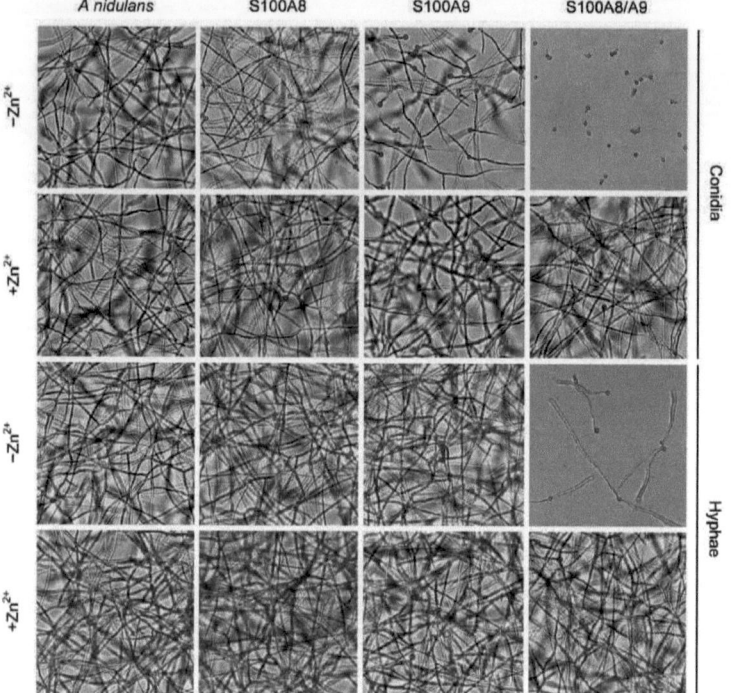

FIG E1. Inhibition of *A nidulans* growth by purified calprotectin is abolished by Zn^{2+} addition. Incubation with the S100A8 (5 μg/mL) subunit did not prevent *A nidulans* conidia or hyphae growth, whereas incubation with S100A9 (5 μg/mL) partially inhibited conidia germination. Coincubation with both subunits strongly inhibited conidia germination and hyphae growth. In all cases, antifungal activity was completely abolished in presence of 1 μmol/L Zn^{2+}.

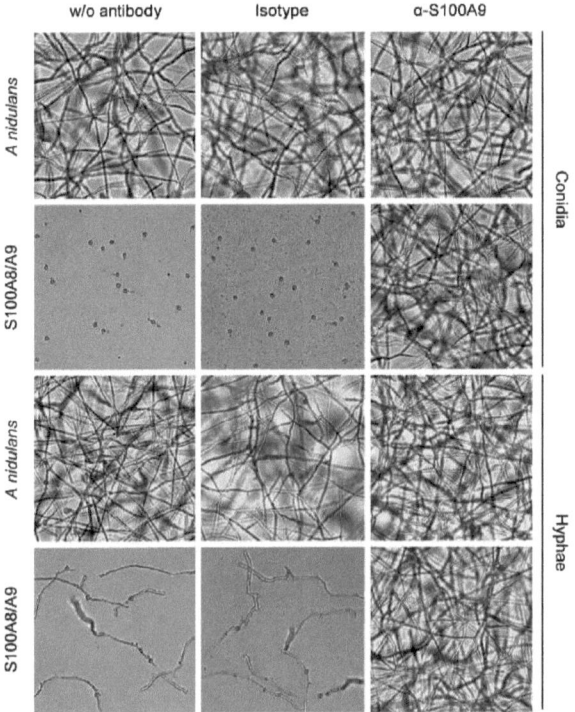

FIG E2. Inhibition of *A nidulans* growth by purified calprotectin is abolished by α-S100A9 antibody. A total of 15 μg/mL polyclonal α-S100A9 antibody fully inhibited antifungal activity of the S100A8/A9 calprotectin complex on conidia or hyphae. As control, S100A8/A9 was incubated with an unspecific control antibody, showing no effect on antifungal activity of the S100A8/A9 calprotectin complex. *w/o*, Without.

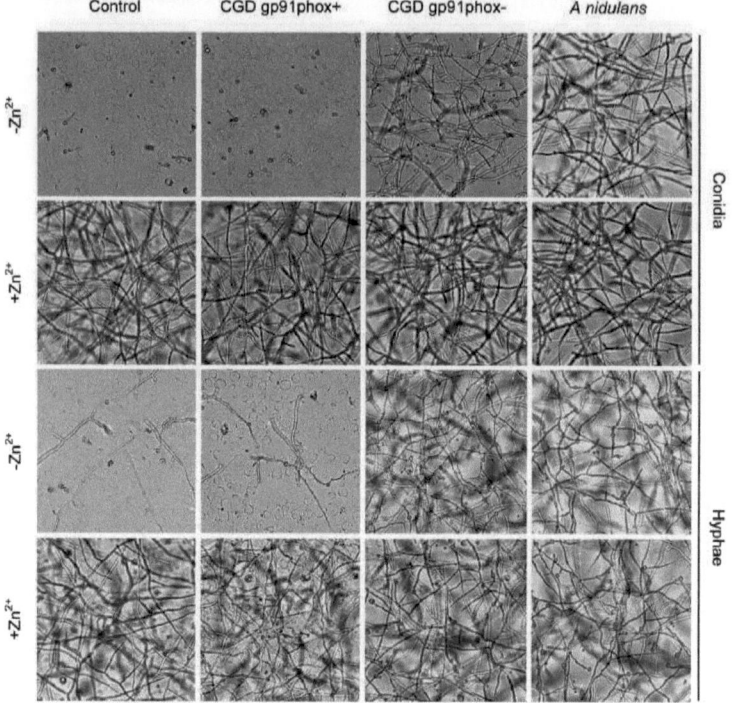

FIG E3. Calprotectin-dependent antifungal activity of CGD gp91[phox+] neutrophils is abolished by Zn^{2+} addition. Control and CGD sorted neutrophils were stimulated with PMA for NET formation and incubated overnight with *A nidulans* conidia or hyphae in presence or absence of 1 μmol/L Zn^{2+}. gp91[phox+] neutrophils showed strong antifungal activity against conidia or hyphae, comparable to control. gp91[phox-] neutrophils were inefficient in controlling fungal growth, with only partial inhibition of conidia germination. In all cases, antifungal activity was completely abolished in the presence of 1 μmol/L Zn^{2+}.

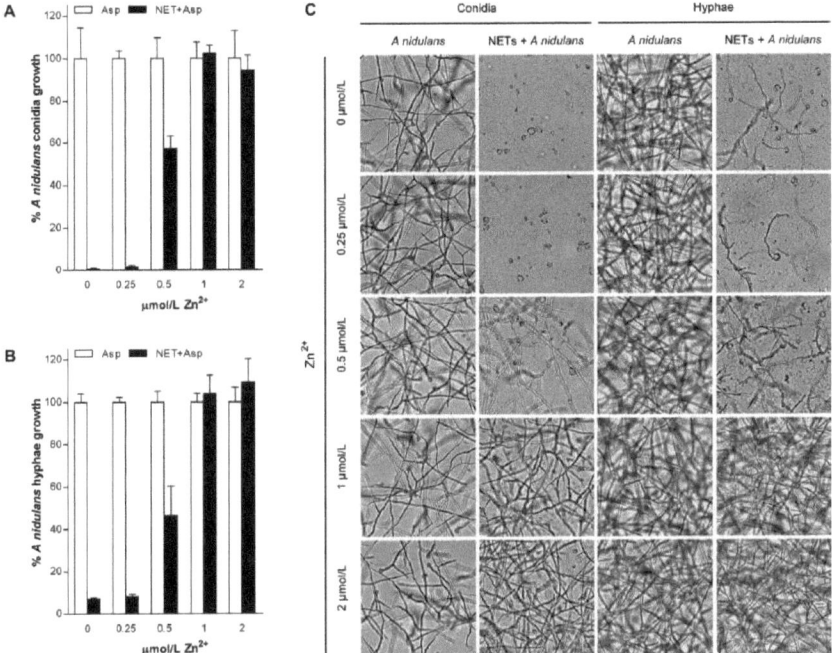

FIG E4. NET inhibition of *A nidulans* growth is abolished by Zn^{2+} addition. Human neutrophils were stimulated with PMA for NET formation and subsequently infected overnight with *A nidulans* conidia or hyphae ± Zn^{2+}. Increasing concentrations of Zn^{2+} abolished the antifungal activity of NETs against conidia **(A)** or hyphae **(B)**. **C**, Representative pictures of data shown in *A* and *B*.

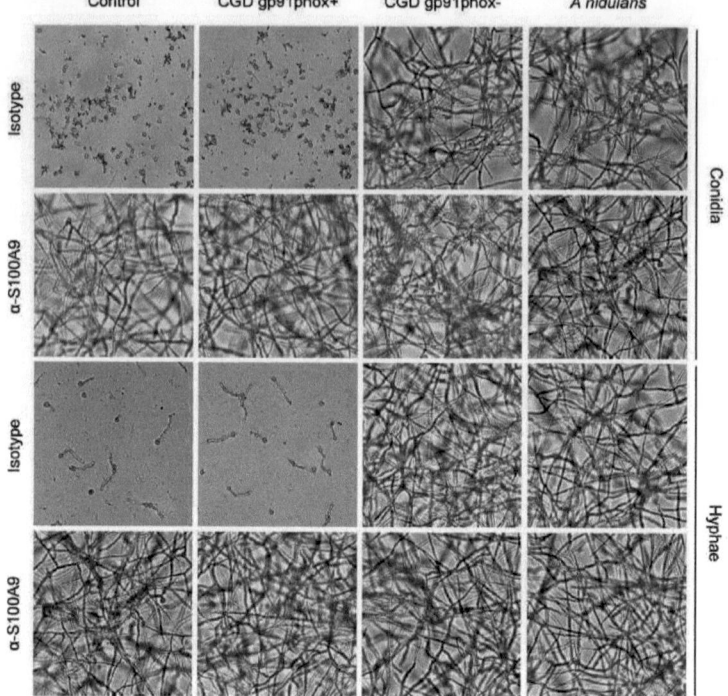

FIG E5. Calprotectin-dependent antifungal activity of CGD gp91^{phox+} neutrophils is abolished by α-S100A9 antibody. Control and CGD sorted neutrophils were stimulated with PMA for NET formation; NET supernatant was digested by DNase/MNase, concentrated, and incubated with *A nidulans* conidia or hyphae after previous incubation with 15 μg/mL polyclonal α-S100A9 or unspecific control antibody. NET extracts from control and gp91^{phox+} neutrophils showed strong antifungal activity against conidia or hyphae. Supernatant from gp91^{phox-} neutrophils had no inhibitory activity on fungal growth. In all cases, fungal growth was completely restored by addition of α-S100A9 antibody, proving that the restored antifungal activity of gp91^{phox+} neutrophils after GT was calprotectin-dependent.

for a short time.[6] Therefore, mammalian neutrophils and/or other phagocytes have most likely redundant abilities to kill *Aspergillus* species that are all ROS-dependent and therefore defective in CGD. However, details remain to be elucidated by appropriate experiments.

Joachim Roesler, MD, PhD
Angela Rösen-Wolff, MD, PhD

From the Department of Pediatrics, University Hospital Dresden, Dresden, Germany. E-mail: Roeslerj@rcs.urz.tu-dresden.de.
Disclosure of potential conflict of interest: The authors have declared that they have no conflict of interest.

REFERENCES

1. Bianchi M, Niemiec MJ, Siler U, Urban CF, Reichenbach J. Restoration of anti-*Aspergillus* defense by neutrophil extracellular traps in human chronic granulomatous disease after gene therapy is calprotectin-dependent. J Allergy Clin Immunol 2011;127:1243-52.
2. Lekstrom-Himes JA, Gallin JI. Immunodeficiency diseases caused by defects in phagocytes. N Engl J Med 2000;343:1703-14.
3. Metzler KD, Fuchs TA, Nauseef WM, Reumaux D, Roesler J, Schulze I, et al. Myeloperoxidase is required for neutrophil extracellular trap formation: implications for innate immunity. Blood 2011;117:953-9.
4. Diamond RD, Clark RA. Damage to *Aspergillus fumigatus* and *Rhizopus oryzae* hyphae by oxidative and nonoxidative microbicidal products of human neutrophils in vitro. Infect Immun 1982;38:487-95.
5. Aratani Y, Kura F, Watanabe H, Akagawa H, Takano Y, Suzuki K, et al. Differential host susceptibility to pulmonary infections with bacteria and fungi in mice deficient in myeloperoxidase. J Infect Dis 2000;182:1276-9.
6. Aratani Y, Kura F, Watanabe H, Akagawa H, Takano Y, Suzuki K, et al. Relative contributions of myeloperoxidase and NADPH-oxidase to the early host defense against pulmonary infections with *Candida albicans* and *Aspergillus fumigatus*. Med Mycol 2002;40:557-63.

Available online June 20, 2011.
doi:10.1016/j.jaci.2011.05.018

Reply

To the Editor:

We thank Roesler and Rösen-Wolff for their comments[1] regarding our recent publication titled "Restoration of anti-*Aspergillus* defense by neutrophil extracellular traps in human chronic granulomatous disease after gene therapy is calprotectin-dependent" in the *Journal of Allergy and Clinical Immunology*.[2] They raise several interesting points, questioning our suggestion that formation of neutrophil extracellular traps (NETs) and NETs-mediated killing of *Aspergillus* species might contribute to clearance in healthy individuals. We concluded this, because neutrophils from patients with chronic granulomatous disease (CGD) do not release NETs on contact with *Aspergillus* and as a consequence killing is impaired. In the following text, we discuss the comments of Roesler and Rösen-Wolff.

Data on myeloperoxidase (MPO) deficiency (MIM #254600) were first published in 1966 by the name of Alius-Grignaschi anomaly, which was defined as total hereditary peroxidase deficiency of neutrophils and monocytes in a healthy person in this publication. Epidemiological data on this rare disease are scarce and controversial (reference list available on request) —incidence rates in the United States and Europe range from 1 case in 2,000 (partial MPO deficiency) to 1 in 4,000 (complete MPO deficiency) population and from 1 in 17,500 (partial MPO deficiency) to 1 in 57,135 (complete MPO deficiency) population in Japan. Some studies do not differentiate between partial and complete deficiency, whereas others report results only from screening of a small collective originating from a restricted geographical area; in addition, a uniform definition of complete MPO deficiency is lacking (eg, null mutation with the absence of protein expression vs hypomorphic mutation with impaired enzymatic function). Information on clinical consequences of complete MPO deficiency, especially with regard to infection, is rare, and a review is lacking to date (reference list available on request). Indeed, a more definitive epidemiological survey would be an important contribution to understanding MPO deficiency.

Therefore, it is difficult to draw any conclusion as to susceptibility to infection of patients completely deficient in MPO. In addition, exposure to infectious agents might vary from one geographical region to another. As Roesler and Rösen-Wolff state correctly, *Candida* infections seem common in patients completely deficient in MPO[3]; however, disseminated infection with *Aspergillus flavus* and pulmonary infection with *Aspergillus fumigatus*[3,4] as well as invasive bacterial infection with *Staphylococcus aureus* and *Legionella* have been described (references available on request). Thus, the increased susceptibility to infection might be linked to the absence of NETs formation in individuals completely deficient in MPO and in patients with CGD,[2,3] although definitive *in vivo* evidence is obviously lacking.

Given the inability of completely MPO-deficient neutrophils to form NETs and some published evidence of susceptibility to infection, identification and a detailed clinical workup of further patients with complete MPO deficiency seems to be warranted in order to clarify its status as a primary immunodeficiency according to the WHO classification (Primary immunodeficiency diseases: an update from the International Union of Immunological Societies Primary Immunodeficiency Diseases Classification Committee Meeting in Budapest, 2005). Such a study would need to analyze genotype-phenotype correlation and susceptibility to infection and optimally comprise *in vitro* analyses of NETs formation on stimulation with identified disease-relevant pathogens.

Susceptibility in murine models should be interpreted carefully, as clinical phenotypes in mice and men do not always correlate. Interestingly, earlier studies with human phagocytes showed that MPO-deficient as well as CGD neutrophils damage *Aspergillus fumigatus* hyphae equally inefficiently.[5,6] Notably, hydrogen peroxide (a product of the nicotinamide adenine dinucleotide phosphate [NADPH] oxidase) or hypochlorous acid (a product of MPO) alone damaged *Aspergillus* hyphae only in concentrations ≥ 1 mM,[5] and extracellular products of MPO did not rescue NETs formation in response to *Candida albicans* stimulation in patients completely deficient in MPO,[3] suggesting that MPO acts cell-intrinsically, independent of its enzymatic activity, most likely by driving chromatin decondensation during NETs formation.[7]

To the best of our knowledge, NETs formation and NETs killing of *Aspergillus* species conidia or hyphae have not been analyzed yet in patients completely deficient in MPO. To address Roesler's and Rösen-Wolff's concerns, neutrophils of patients completely deficient in MPO should be analyzed parallely with those of patients with CGD and healthy controls and by adding products of MPO (eg, hypochlorous acid, histamine chloramines) and/or NADPH oxidase (eg, H_2O_2) to neutrophils.

We therefore conclude that according to current knowledge, both NADPH oxidase and MPO are required for fungus-induced NETs formation. However, the apparent divergence of major pathogens related to CGD and complete MPO deficiency, concluded from scarce, thus potentially unreliable, epidemiological data in complete MPO deficiency, remains to be defined.

Matteo Bianchi, MSc[a]
Maria J. Niemiec, MSc[b]
Ulrich Siler, PhD[a]
Constantin F. Urban, PhD[b]
Janine Reichenbach, MD[a]

From [a]the Division of Immunology/Hematology/BMT, University Children's Hospital, Zurich, Switzerland, and [b]the Antifungal Immunity Group, Molecular Biology Department, Laboratory for Molecular Infection Medicine Sweden, Umeå University, Umeå, Sweden. E-mail: janine.reichenbach@kispi.uzh.ch.
Disclosure of potential conflict of interest: J. Reichenbach and M. Bianchi have received research support from the Chronic Granulomatous Disorders (CGD) Research Trust. The rest of the authors have declared that they have no conflict of interest.

REFERENCES

1. Roesler J, Rösen-Wolff A. Redundant ability of phagocytes to kill *Aspergillus* species. J Allergy Clin Immunol 2011;128:686-7.
2. Bianchi M, Niemiec MJ, Siler U, Urban CF, Reichenbach J. Restoration of anti-*Aspergillus* defense by neutrophil extracellular traps in human chronic granulomatous disease after gene therapy is calprotectin-dependent. J Allergy Clin Immunol 2011;127:1243-52.
3. Metzler KD, Fuchs TA, Nauseef WM, Reumaux D, Roesler J, Schulze I, et al. Myeloperoxidase is required for neutrophil extracellular trap formation: implications for innate immunity. Blood 2011;117:953-9.
4. Chiang AK, Chan GC, Ma SK, Ng YK, Ha SY, Lau YL. Disseminated fungal infection associated with myeloperoxidase deficiency in a premature neonate. Pediatr Infect Dis J 2000;19:1027-9.
5. Diamond RD, Clark RA. Damage to *Aspergillus fumigatus* and *Rhizopus oryzae* hyphae by oxidative and nonoxidative microbicidal products of human neutrophils in vitro. Infect Immun 1982;38:487-95.
6. Rex JH, Bennett JE, Gallin JI, Malech HL, Melnick DA. Normal and deficient neutrophils can cooperate to damage *Aspergillus fumigatus* hyphae. J Infect Dis 1990; 162:523-8.
7. Papayannopoulos V, Metzler KD, Hakkim A, Zychlinsky A. Neutrophil elastase and myeloperoxidase regulate the formation of neutrophil extracellular traps. J Cell Biol 2010;191:677-91.

Available online June 20, 2011.
doi:10.1016/j.jaci.2011.05.019

Effect of partially hydrolyzed whey infant formula and prolonged breast-feeding on the risk of allergic disease in high-risk children

To the Editor:

We reviewed with interest the article by Lowe et al.[1] The authors conclude the following: "Despite current dietary guidelines, we found no evidence to support recommending the use of pHWF [partially hydrolyzed whey infant formula] at weaning for the prevention of allergic disease in high-risk infants." As a sponsor of this study, Nestlé R&D feels obligated to make the following comments.

A 2006 Cochrane Review[2] on hydrolyzed formulas and allergy prevention assessed an unpublished 2-year report from the authors and excluded this trial from analysis because of "excess randomisation losses 238/620 (38%)." Nestlé R&D also assessed this report,[3] which stated that 238 infants received a "nonassigned" formula. This differs drastically from the 26 infants listed in the current article. This degree of nonadherence to formula allocation implies that a very large number of infants in all groups were exposed to intact proteins. This seriously jeopardizes any intervention aimed at reducing the risk of allergy with a hydrolyzed formula, and no statistical analysis can overcome this shortcoming.

In the current publication the first 97 infants were randomized to 2 formulas before the third one became available. In addition, 50% of subjects by 4 months and 39% by 6 months of age were still breast-feeding and did not receive their allocated formula. It is excellent that recommendations of prolonged breast-feeding were followed. In fact, the results could reflect a potential protective effect of breast-feeding in the whole study cohort. However, the low exposure to allocated formula coupled with prolonged breast-feeding introduces considerable bias, significantly decreasing the chances for identifying a difference caused by the formulas.

All clinical outcome measures were assessed through parental "telephone interview." No accepted diagnostic criteria for atopic dermatitis were used. There was no documentation of allergic disease between 2 and 6 years of age, leaving 4 of 7 years of follow-up undocumented. Also, documentation of allergic conditions for 6 to 7 years was done with a single telephone call at "6 or 7" years of age.

In summary, Nestlé R&D has significant concerns regarding the conclusions of this study, which is being published 21 years after its initiation and which, among other limitations, was single-blind, lacked strict diagnostic criteria, had high noncompliance rates, and had gaps in subject follow-up.

On the basis of a large body of evidence that includes studies of high quality, a number of professional organizations and expert groups[4,5] have concluded that use of certain hydrolyzed infant formulas, in particular partially hydrolyzed whey infant formula, has a role in the reduction of the risk of atopic disease, particularly atopic dermatitis. Also, 2 recent meta-analyses[6,7] confirm this benefit.

Safe alternatives to reduce allergic risk associated with intact cow's milk protein formulas are increasingly needed for infants who do not receive all the benefits of exclusive breast-feeding. The conclusions of a single study with the limitations mentioned above is of little consequence and should not affect current recommendations, which are in the interest of advancing infant health.

Ferdinand Haschke, MD, PhD

From the Nestlé Nutrition Institute, Vevey, Switzerland. E-mail: Ferdinand.haschke@nestle.com.
Disclosure of potential conflict of interest: F. Haschke is Chairman of the Nestlé Nutrition Institute, a separate non–profit-making legal entity at Nestlé, which is active in the field of medical/scientific communication. He is employed by the Nestlé Group, from which he receives a salary.

REFERENCES

1. Lowe AJ, Hosking CS, Bennett CM, Allen KJ, Axelrad C, Carlin JB, et al. Effect of a partially hydrolyzed whey infant formula at weaning on the risk of allergic disease in high-risk children: a randomized controlled trial. J Allergy Clin Immunol 2011;128: 360-5.e4.
2. Osborn DA, Sinn J. Formulas containing hydrolysed protein for prevention of allergy and food intolerance in infants. Cochrane Database Syst Rev 2006;(4): CD003664.
3. Hill D. A report on the analysis of the Melbourne Atopy Cohort Study. A study designed to test the effectiveness of different formula types on the development of atopic symptoms and signs on a cohort of atopy-at-risk infants. Report at year 2. Nestec internal report. Lausanne (Switzerland): Nestec; 1996.
4. Host A, Koletzko B, Dreborg S, Muraro A, Wahn U, Aggett P, et al. Dietary products used in infants for treatment and prevention of food allergy: joint statement of the European Society for Paediatric Allergology and Clinical Immunology (ESPACI) Committee on Hypoallergenic Formulas and the European Society for Paediatric Gastroenterology, Hepatology and Nutrition (ESPGHAN) Committee on Nutrition. Arch Dis Child 1999;81:80-4.
5. Guidelines for the diagnosis and management of food allergy in the United States: report of the NIAID-sponsored expert panel. J Allergy Clin Immunol 2010; 126(suppl):S1-58.

Acknowledgements

It is a pleasure to thank those who made this thesis possible:

PD Dr. med. Janine Reichenbach, the supervisor of my thesis, first for giving me the opportunity to perform my PhD at the University Children's Hospital Zurich, then especially for her ideas, support and positiveness during all the project. Dr. Ulrich Siler, for all his advices and help during lab and office work. Prof. Dr. Reinhard A. Seger, head of the Immunology/Hematology/BMT division, for sharing his knowledge and enthusiasm about CGD and research.

I wish to thank Prof. Dr. Cornel Fraefel for agreeing to chair the PhD-committee and for helpful discussions. Prof. Dr. Alexandra Trkola and Prof. Dr. Michael Hengartner for attending the committee and for all their suggestions.

A great thank to all collaborators for their trust, help and exciting scientific discussions: Prof. Dr. Arturo Zychlinsky and Abdul Hakkim (Max Planck Institute for Infection Biology, Berlin, Germany); Dr. Constantin F. Urban and Maria J. Niemiec (Umeå University, Umeå, Sweden); Dr. Manuel Grez (Georg Speyer Haus, Frankfurt, Germany).

Furthermore, I am grateful to all my past and present lab colleagues: Vital Wohlgensinger, Elke Karaus, Andrea Valero, Sabine Muff, Julie Lemay, Walther Hänseler, Elena Kouzmenko and Katarzyna Nytko, for their help and pleasant time inside and outside of the lab.

I would also like to thank the Immunology/Allergy lab members: Remo Frei, Caroline Roduit, Susanne Löliger and Johanna Wohlgensinger, for their help and great time spent together.

A last thank to all people and neighboring labs I bothered during my thesis, and especially to Imane Azouzzi, Martin Stucki and Michele Frapolli for the *ex lab* fun we had together.

Finally, I am indebted to my family. Without them whatever happened in the last 30 years would not have been possible.

Acknowledgements

The presented work was supported by a research grant of the University of Zurich, Switzerland (Forschungskredit der Universität Zürich, Kreditnummer 54191402) and by a grant of the Chronic Granulomatous Disorder Research Trust, United Kingdom (grant no. J4G/08/01).

  Challenging genetic disorders
The CGD Research Trust

Curriculum Vitae

Name	**Matteo BIANCHI**	
Date of birth	07.02.1981 in Sorengo (TI), Switzerland	
Citizenship	Cureglia (Ti), Switzerland	
Nationality	Swiss	

Education

1987-1992	Primary school, Cureglia, Switzerland
1992-1996	Secondary school, Savosa, Switzerland
1996-2000	High school Lugano 2, Diploma type C (scientific), Lugano, Switzerland
2000-2005	Studies of Biology at the University of Lausanne, Switzerland
2000-2002	First cycle of Biology studies
2002-2003	Certificate in Molecular and Cellular Biology
2003	Certificate in Zoology
2003-2004	Certificate in Biochemistry
2004	Certificate in Microbiology
2004-2005	Diploma Thesis: "Vaccin thérapeutique contre HPV et status inflammatoire du tractus génital de la souris" (Therapeutic vaccine against human papilloma virus and inflammatory status of the murine genital tract) Performed in Prof. Denise Nardelli-Haefliger's group, Department of Gynecology, CHUV, Lausanne
2006-2011	Employed as PhD student at the University of Zurich, in the field of gene therapy for chronic granulomatous disease and innate immunity against fungal infections at the University Children's Hospital Zurich, Switzerland Thesis performed in Prof. Reinhard Seger's group, under the supervision of PD Dr. med. Janine Reichenbach, Division of Immunology/Hematology/BMT, Jeffrey Modell Diagnostic and Research Center for Primary Immunodeficiencies, University Children's Hospital Zurich, Switzerland

Professional education during doctorate studies

2007	Course for animal experimentation, University of Lausanne
2009-2010	Course of scientific writing and publishing, University of Zurich
2010	Course for generation of transgenic animals, University of Zurich

Publications during doctorate studies

Bianchi M, Niemiec MJ, Siler U, Urban CF, Reichenbach J. Correspondence reply: *Aspergillus* infection in CGD and MPO deficiency. *J Allergy Clin Immunol* 2011;128:687-688. **Impact factor: 9.273**

Bianchi M, Niemiec MJ, Siler U, Urban CF, Reichenbach J. Restoration of anti-*Aspergillus* defense by neutrophil extracellular traps in human chronic granulomatous disease after gene therapy is calprotectin-dependent. *J Allergy Clin Immunol* 2011;127:1243-1252. **Impact factor: 9.273**

Hakkim A, Hurwitz R, **Bianchi M**, Brinkmann V, Siler U, Seger RA, Zychlinsky A, Reichenbach J. Correspondence reply: Protecting against *Aspergillus* infection in CGD. *Blood* 2009;114:3498. **Impact factor: 10.558**

Bianchi M, Hakkim A, Brinkmann V, Siler U, Seger RA, Zychlinsky A, Reichenbach J. Restoration of NET formation by gene therapy in CGD controls aspergillosis. *Blood* 2009;114:2619-22. **Impact factor: 10.558**

Echchannaoui H, **Bianchi M**, Baud D, Bobst M, Stehle JC, Stamenkovich I, Nardelli-Haefliger D. Intravaginal Immunization of mice with recombinant *Salmonellae* expressing human papillomavirus type 16 antigens as a potential route of vaccination against cervical cancer. *Infection and Immunity* 2008;76:1940-51. **Impact factor: 4.098**

Congresses attended during doctorate studies (active participation – posters)

2009	17th European Society of Gene & Cell Therapy congress (ESGCT), Hannover (D)
2010	9th International Mycological Congress (IMC9), Edinburgh (UK)
2010	9th day of clinical research, Zurich (CH)
2011	98th AAI Immunology Annual Meeting, San Francisco (USA)
2011	10th day of clinical research, Zurich (CH)

Grants received during doctorate studies

2006-2008	PhD grant from the University of Zurich (CH) (Forschungskredit der UZH)
2009-2011	PhD grant from The CGD Research Trust Foundation (UK)

Awards

2011	10th day of clinical research, University Hospital Zurich (CH) – Best Poster Award

i want morebooks!

Buy your books fast and straightforward online - at one of world's fastest growing online book stores! Environmentally sound due to Print-on-Demand technologies.

Buy your books online at
www.get-morebooks.com

Kaufen Sie Ihre Bücher schnell und unkompliziert online – auf einer der am schnellsten wachsenden Buchhandelsplattformen weltweit! Dank Print-On-Demand umwelt- und ressourcenschonend produziert.

Bücher schneller online kaufen
www.morebooks.de

VDM Verlagsservicegesellschaft mbH
Heinrich-Böcking-Str. 6-8　　Telefon: +49 681 3720 174　　info@vdm-vsg.de
D - 66121 Saarbrücken　　　Telefax: +49 681 3720 1749　　www.vdm-vsg.de

Printed by Books on Demand GmbH, Norderstedt / Germany